Applied Ergonomics

Applied Ergonomics

Edited by

David C. Alexander

and

Randall A. Rabourn

CRC Press
Taylor & Francis Group
Boca Raton London New York

CRC Press is an imprint of the
Taylor & Francis Group, an **informa** business
A TAYLOR & FRANCIS BOOK

First published 2001 by Taylor & Francis

Published 2019 by CRC Press
Taylor & Francis Group
6000 Broken Sound Parkway NW, Suite 300
Boca Raton, FL 33487-2742

First issued in paperback 2019

No claim to original U.S. Government works

ISBN-13: 978-0-367-45523-1 (pbk)
ISBN-13: 978-0-415-23852-6 (hbk)

**Visit the Taylor & Francis Web site at
http://www.taylorandfrancis.com**

**and the CRC Press Web site at
http://www.crcpress.com**

Publisher's Note
This book has been prepared from camera-ready copy supplied by the authors

Every effort has been made to ensure that the advice and information in this book is true and accurate at the time of going to press. However, neither the publisher nor the authors can accept any legal responsibility or liability for any errors or omissions that may be made. In the case of drug administration, any medical procedure or the use of technical equipment mentioned within this book, you are strongly advised to consult the manufacturer's guidelines.

British Library Cataloguing in Publication Data
A catalogue record for this book is available from the British Library

Library of Congress Cataloging in Publication Data
A catalogue record for this book has been requested

Contents

SUPPORT AND SERVICE JOBS

OFFICE ENVIRONMENTS

Introduction

David C. Alexander, PE, CPE
Randall A. Rabourn, CSP, CPE

Applied Ergonomics is a text devoted to sharing the experiences of ergonomics practitioners in a variety of applications, situations and environments. The chapters of this text are a series of papers authored by practicing ergonomists to provide an insight into the "real world" application of ergonomics principles. Each chapter represents a paper presented at the Third Annual Applied Ergonomics Conference in Los Angeles, California, March, 2000. A heartfelt "thanks" is extended to the individual authors for their willingness and effort to document their experiences so that other ergonomics specialists can benefit.

There are thirty-five chapters in this text and they are organized into sections by the type of ergonomics application. The application sections include: manufacturing and production processes; support and service jobs; office environments; potpourri; health management; and design. Each chapter can be read in the authors' own words as they share what has worked, or not worked, for them. Each chapter will have relevance to you based upon your own work, your environment, and your degree of experience and formal ergonomics training. Nonetheless, you should find all of the chapters valuable in helping you decide appropriate actions to take in order to improve ergonomics workplace conditions.

MANUFACTURING AND PRODUCTION PROCESSES

Manufacturing and production process environments continue to be the primary areas for ergonomics applications, and a proportionately large part of this book is devoted to examples from these areas. Chapters 1 through 15 comprise this section and address a range of topics, from company ergonomics process and program issues to ergonomic intervention case studies.

A series of papers describes ergonomics processes and programs developed and applied by several companies. The focus of "An Ergonomics Lineside Audit Method in Automotive Manufacturing" by Smith, is an ergonomics audit procedure used by an automotive manufacturer. This audit activity is described in detail, and is the first step in the company's goal to develop an "ergonomics culture." The following paper by Henderson, "Textron Automotive Trim: Initiating a Sustainable Ergonomics Process," outlines key elements of an ergonomics process at an automotive trim plant and describes the steady reduction of musculoskeletal disorders attributed to adhering to this process over a number of years. The next paper, "Ergonomics Before Equipment Receipt," by White and Schwab, describes the process one company uses to ensure its equipment vendors adhere to industry ergonomics specifications for manufacturing equipment design.

The ability to identify and solve ergonomics problems simultaneously for many production facilities in a multi-plant system presents significant challenges. "Sharing Ergonomics Solutions in a 'Virtual Factory' System at Intel Corporation" by House and Harrison describes the "virtual factory" process employed by a large multi-national corporation to consistently and efficiently address common ergonomics issues at all of its plants. Another systematic approach for ergonomic change is described in "Deferential Analysis and Integrated Solutions" by Povlotsky, Georgoff and Malone. This paper illustrates how information from many sources and stakeholders is integrated to develop feasible and acceptable ergonomics solutions. These papers present innovative methods to affect ergonomic change.

Written company ergonomics plans demonstrate management commitment to ergonomics, provide a systematic plan or "road map" to handle ergonomics issues, and provide documentation that ergonomics issues are being addressed. Howell's paper, "Whirlpool Corporation Clyde Division Ergonomic Process Management Written Plan," lists some key elements of the ergonomics program in an appliance manufacturing division and includes the actual written plan. People interested in developing or evaluating their own written plans will find this example valuable.

The remaining nine papers in this section are grounded in case studies. Cases range from the design of entire work systems involving many jobs and workers to the development of specialized tools to eliminate job-specific ergonomic risk factors. Adams' "Using Human Factors To Understand and Improve Machine Safeguarding" describes how ergonomic design strategies can be used for positively affecting workers' safety behavior. His paper focuses on machine safeguarding issues and provides examples that the ergonomist can follow to encourage the proper use of machine safeguarding systems.

"Overcoming Traditional Job Rotation Obstacles in the Design of a Manufacturing Cell" by Wyatt describes the design of two production cell systems involving several workers. Practical issues of worker job rotation within cells are discussed, and ergonomic and economic aspects of the cell designs are compared. The following paper by Davis, "High-Tech but Low-Cost Ergonomic Solutions," looks at the need to employ emerging technologies to address ergonomics problems not adequately solved by conventional methods. Examples from three different industries are described along with a discussion of when "high-tech" solutions should be explored.

A categorical overview of materials handling aids is provided in Chuckrow's "From Handtrucks to Palletizers: Advances in Material Handling Equipment for the New Millennium." His paper will be a resource for readers who are selecting equipment to reduce stress and improve productivity in many manual materials handling situations.

The remaining case studies are examples of interventions and intervention strategies to address specific ergonomics problems. Ergonomic solutions and solution development processes are discussed. Brewer's paper, "Development & Implementation for Two Specialized Material Handling Devices: Metal Plate Movement & Oven Loading for Precision Optics," describes two ergonomic interventions that improved production operations while simultaneously reducing ergonomic stressors. The need to evaluate and "fine tune" interventions during the implementation phase is also discussed in his paper.

Meat and poultry processing operations face many ergonomics challenges and have been the focus of considerable attention over the past few years. Strides have been made to improve the ergonomic conditions in many of these operations. Ergonomic interventions implemented by several different processors are described in the papers by DeRoos ("Pork Loin Processing Lines"), Worrell ("Development of a Customized Ergonomic Handle: Meat Industry Application" and "Elimination of the Incidence of CTDs from the Traditionally Top Risk Job in Beef Meatpacking"), and Ramcharan ("Ergonomic Redesign of a Hanging Table for Poultry Processing"). Ergonomically sound tools and devices designed specifically for meat and poultry processing tasks are illustrated along with approaches that these practitioners used to evaluate these interventions. Lessons from these industry-specific papers should be transferable to other production operations.

SUPPORT AND SERVICE JOBS

The good news is that ergonomics is being used in more diverse settings. As ergonomics is becoming commonplace in industrial and office settings, it is also making its entry into other support and service areas. As with any new and novel applications of ergonomics, the first requirement is to simply recognize that an ergonomics problem exists. The next step, then, is to resolve that problem and sell its solution to both management and workers. These are not easy tasks. Recognizing ergonomics problems is easier in industrial and office setting when those same problems have been identified and discussed in the past. Developing new and novel solutions, or even adapting existing solutions to a new environment, is always challenging. And finally, selling and implementing the solution, in places where management and workers are still "warming up" to ergonomics is also challenging.

Yet, Chapters 16 through 19 provide four examples of successful projects in the support and service environment. The first, "Bottled Water Handling: A Case Study Through Biomechanical Analysis" by Cotnam, Chang and Garrison, reports on the task of delivering bottled water and provides recommendations for manual handling of bottled water. As important, it reports on methods that are less safe, thus permitting the best work methods to be used immediately.

The next paper, "Work-Related Musculoskeletal Disorders: Ergonomic Risk to Healthcare Workers" by Vredenburgh and Zackowitz, discusses the problems facing health care workers, and provides many recommendations for the control of their ergonomics problems. The solutions cover patient handling, furniture and equipment movement, trash and laundry management, and even floor care. This article is remarkable in its quantity, variety and depth of solutions.

The next paper, "Ergonomic Solutions in Electric Energy Generation" by Barracca, describes the author's ergonomics work in electric energy generation. With this paper, he describes the ergonomics problems facing maintenance workers, and provides descriptions of a series of projects that evaluate the ergonomics stresses on these workers. These projects resulted in specific detailed recommendations for the use of gloves, knee protection, hammers, material handling devices, and tool balancers.

The final paper in this section, "Making Ergonomic Changes in Construction: Worksite Training and Task Interventions" by Hecker, Gibbons, and Barsotti, describes a demonstration project used to identify ergonomics risks encountered by various construction trades. The work was performed at a common construction site, and the work was performed with a general contractor and various sub-contractors. In addition to training, interventions were made in utilities installation, concrete core drilling, pipe cutting and reaming, threading steel rods, and installing air filters and blanks. The concepts used in these interventions have value beyond the initial application.

OFFICE ENVIRONMENTS

Applications in office ergonomics continue to evolve, and this series of papers in Chapters 20 through 24 provides new insight into office ergonomics solutions. The first paper, "Visual Ergonomics in the Workplace" by Anshel begins with a review of the eyes as the visual system, then moves on to describe computer vision syndrome. Issues of workplace lighting including color and the quality and quantity of light are presented. General eye care tips are provided to help the computer user comfortably and safely work in the office environment.

The second paper in this section, "Office Ergonomics Assessments Using the Internet" by Kestler and Romero, describes a methodology that permits office workplace evaluations using the internet. The results are dramatic with high satisfaction and accuracy ratings, and a much lower cost than traditional on-site evaluation methods. The process used for the evaluations is carefully and clearly outlined as is methodology used to evaluate the new technique.

The next paper, "ERGO NIGHTMARE and Other Adventures in Starting a Grassroots Office Ergonomics Program" by Sisler provides a humorous and insightful review of the experiences involved in initiating and managing an ergonomics program. In addition to implementing sound ergonomics practices (training, equipment modifications, work/recovery regimens, eye care, etc.), this group also had fun with the implementation. It was clearly a win-win program for both the program leaders and the computer users.

The next chapter, "Questioning Office Ergonomic Guidelines" by Ankrum, invites us to question and understand the multitude of office ergonomics guidelines found in the marketplace. With compelling examples, the author challenges us to look carefully at work posture and dimensional guidelines for office workplaces and ensure that they provide the safety and ergonomic protection intended. Conventional wisdom and traditional recommendations may not always provide the best recommendations.

The final paper in this section, "Resolving RMI's in Manual Insurance Policy Assembly" by Borretti, describes a traditional office task and some ergonomics alternatives to its resolution. The task is manual assembly, similar to a job shop assembly task in a product-based industry. The solutions follow a traditional path and the process used to identify, evaluate, and implement the solutions is explained well.

POTPOURRI

With any collection of papers on ergonomics applications, there are always several intriguing and insightful articles that don't fall within traditional groupings or categories. Chapters 25 through 29 contain these papers and the delights associated with finding such new and novel information and approaches to ergonomics justification, training, program management, and certification.

The first paper provides an approach and examples of cost justification of ergonomics. This is an important area and one which many ergonomics practitioners have a strong interest. In their article, "The Economics of Ergonomics: Three Workplace Design Case Studies," Jenkins and Rickards discuss their philosophy of cost justification and present three case studies with dramatic rates of return. These are excellent examples that demonstrate the economic value of sound ergonomics projects.

The next chapter also deals with cost justification. Vaidya and Weeks, in their paper "Using NCCI Rating Worksheet in Cost/Benefit Analysis," explain some of the intricacies of the insurance industry and how these can be helpful in cost/benefit analysis. They also provide an example to illustrate their process. This article is a "must read" for anyone performing cost-benefit analysis.

The third paper in this section, "The Power Zone," by Hilgen, takes complex information and turns it into a simple concept—the power zone. The power zone is the area in which the employee can work easily and safely, and that each individual can easily identify. It can be quickly taught to workers, and is readily observed. Several examples are given for workers in tree care, landscaping, dry wall, and manufacturing.

The fourth paper shifts to a look at ergonomics programs and what makes them effective. Vredenburgh and Zackowitz present their ideas in "Ergonomics Programs: Reducing Work-Related Musculoskeletal Injuries." In this paper, they describe an effective ergonomics program, and provide nine steps to implement that program. They use an example to knit their concepts together, and effectively explain ergonomics program management and implementation.

The final paper in this section, "Ergonomics Certification: Its Value and Quality," by Brauer and Rice discusses certification for ergonomics professionals. This chapter, written by representatives of two leading certification groups, is a thorough and careful review of certification, the processes used for certification, and the outcomes. It is essential reading for individuals who consider themselves to be professional ergonomists.

HEALTH MANAGEMENT

Health management and the care of injured workers remain an important part of any ergonomics program. Perhaps even more important than healing an injury is the prevention of that injury. Chapters 30 and 31 address these issues.

In Chapter 30, "Job Simulation and Physical Abilities Testing," Valentine and Albarino describe how they have developed and utilized physical abilities testing to ensure that workers can perform the jobs for which they are assigned. The test equipment and the test protocol are described. Initial results are favorable.

In the next paper, "Nerve Conduction Testing: Post-Offer Job Assessments," Johnson describes his process and results for nerve conduction testing in the context of the employment process. Testing includes physical examination, standard non-invasive tests, and nerve conduction testing. This is a good overview of the practical as well as the scientific aspects of post-offer testing processes.

DESIGN

Design for ergonomics offers great promise for the prevention of ergonomics problems and injuries. Currently, ergonomics design is still an emerging aspect of ergonomics and it is wonderful to be able to share so many different examples of ergonomics design in this text. Design processes and tools, as well as case studies are provided in Chapters 32 through 35. These design examples span a wide range from facility design, to equipment design, to tool design, and even product design.

In the first paper, "An Ergonomic Approach to Facility Design and Layout," Brown presents his process for the ergonomics and human factors design of a medical clinic. He describes his process for design, and then illustrates the process with a very well documented example. The issues faced are novel, yet he addresses them in a practical and easily understood manner.

The next paper, "Design of a Hoist for the Removal and Replacement of Horizontal Semiconductor Wafer Furnace Heater," Meekins describes a classic situation involving a critical task that has severe ergonomic limitations, all coupled with extensive facility constraints. The process of design is described, alternatives are considered and improved, and finally, a working solution is obtained. This is a very good case study showing the blood, sweat and tears of a real design example.

The next paper is a tool design case study. Johnson and Wasserman present the story of a "Powered Circular Knife to Reduce Worker Fatigue." The important issues are grip and control, involving the tool materials and vibration. Testing done in the workplace is a final part of the overall evaluation.

The final chapter involves product design. In this case Benden describes the classic situation of turning a lemon into lemonade. The article is entitled "Use of Three Dimensional Fabric Yields/Ergonomics Benefits in Upholstery Work." He begins with a description of a problem placing fabric covers on an office chair. While exploring options for improving the task, a new fabric material is considered. The new fabric not only eases the work and controls the injury potential, it also provides more comfort for the user and is faster to assemble.

MANUFACTURING AND PRODUCTION PROCESSES

CHAPTER ONE

An Ergonomics Lineside Audit Method in Automotive Manufacturing

Robert T. Smith, Ph.D., CIE

1.1 INTRODUCTION

Honda of America Mfg., Inc. (HAM) has long recognized the importance of ergonomics in its manufacturing operations. In order to enhance our existing expert-based ergonomics program we decided to pursue the development and implementation of an "Ergonomics Culture." This Ergonomics Culture concept is one of moving from an "Expert help" model to "Skilled help," and eventually moving seamlessly to "Self help," all within the company.

A common theme in many ergonomics programs is to rely entirely on company experts to resolve ergonomics problems. The expert help approach stretches a company's often-limited ergonomics personnel resource. In contrast, an established Ergonomics Culture makes recognizing and solving ergonomics problems everyone's job. The intent of this culture is to move from a reactive ergonomics program to a proactive program. Reactive ergonomics relies on lagging indicators and usually identifies problems by recognizing an increase in injuries and mounting workers' compensation costs. Conversely, a proactive ergonomics program anticipates and avoids problems by seeking out leading indicators like manufacturing change points, actively pursuing and participating in design opportunities, and recognizing and then correcting ergonomics risk factors before they cause injury.

We recently conducted a situation analysis at HAM to determine ergonomics program needs, and to learn from past experiences. After reviewing our findings, we began team-building and drafted a long-term plan to achieve a fully implemented Ergonomics Culture. An early step in our plan included an audit of ergonomics systems and lineside jobs (called processes at HAM), in order to determine a baseline of current ergonomics activities and to establish priorities.

1.2 AUTOMOTIVE MANUFACTURING

1.2.1 Manufacturing Operations

This case study discusses extensive ergonomics audits planned and conducted at four large automotive manufacturing facilities. Honda workers in all these facilities are called associates. Manufacturing operations to be audited covered the entire spectrum of automotive production including casting, machining, plastics, parts distribution, stamping, welding, painting, assembly, inspection and shipping.

The automotive assembly functions are especially hand-intensive manual work. Fortunately, Honda automobile products are designed for comparatively easy assembly. Nevertheless, manually applied parts and pieces often require upper-extremity effort and the use of powered tools.

Work assignments or processes at HAM usually constitute a set of consecutive tasks in the car (or motorcycle) production operations. The automotive manufacturing work environment is very dynamic, with at least annual product refinements and periodic major model changes. These manufacturing change points can have a significant impact on how each process is configured and performed. Auditing ergonomics in this setting requires recognizing the manufacturing changes during process modifications, and taking advantage of them as opportunities to introduce ergonomics improvement.

1.2.2 Pre-Audit Preparation

In preparing to conduct audits in any manufacturing environment it is a challenge to avoid or minimize interfering with production and/or quality. In the automotive industry, with its just-in-time inventory and manufacturing processes, and its demanding quality standards, any new activity, such as an audit process, is scrutinized to ensure that production and quality ideals are not compromised.

To satisfy production concerns, a realistic plan and functional audit flow must be developed, presented and revised, to assure consensus with the audit's operations and objectives. Drafting a detailed audit procedural flow before starting, such as precisely who does what, exactly when do they do it and how is it to be done, can avoid some common pitfalls (Kennedy *et al.*, 1992; Keyserling *et al.*, 1993). To meet production management's expectations, a draft of all plans, schedules and procedures relating to the audits should be prepared and presented in a lineside mock audit demonstration. This should be scripted and rehearsed to show clearly to production management that auditors will not be disruptive, and that they will be well managed, while at the same time accomplishing their purpose of auditing.

1.3 AUDIT METHODOLOGIES

For those seeking advice on the ins and outs of ergonomics audit methodologies, Colin Drury (1998), Kirwan and Ainsworth (1992), and McAtamney and Corlett (1993) provide excellent discussions of ergonomics audit methods. There many varied methods to choose from, but care must be exercised in the selection to ensure that the method fits the need and the result is productive. Our approach included observational, interview and questionnaire methods, which required preparation of formats and protocols to establish consistency. Conducting observational task analyses, interviewing production associates while they are adjacent to their workplace, and administrating ergonomics questionnaires for auditing must be accomplished without being intrusive or interrupting production.

Self-examination by auditing can be an uncomfortable prospect for many organizations. To gain consensus and support for our audits, production departments were asked to assign safety and engineering associates to receive basic ergonomics training and to assist consultants in performing the audits. This increased ergonomics awareness among associates, improved their skills and gave audit team associates an opportunity to participate. It also prepared them to understand better the ergonomics concerns they would identify during the audit and would later facilitate implementation of countermeasures. This training supported our long-term training strategy, contributing to the Ergonomics Culture's development.

During planning, we recognized that at least two distinct audits should be conducted: a management "systems" audit, and an audit of the manufacturing workplace or "lineside" audit. Before starting lineside audits, a third-party consultant was retained to conduct a systems audit to assess the state of the ergonomics management systems. This systems audit assessed management's performance and support for ergonomics programs, and the results were reported to senior management. Substantial planning and preparation are needed to conduct any audit, and a systems audit is no exception. However, this case study discusses only the lineside audit method.

1.4 LINESIDE AUDIT METHOD

1.4.1 Method Selection

Selecting an audit method walks the fine line between finding just the big problems, finding ALL the problems or maybe finding none. In the ideal world the ergonomist's inclination is to look for all of the ergonomics risk factors. However, we intended to establish a program baseline, so identifying significant ergonomics concerns provided a good foundation (Kuorinka and Forcier, 1995; NIOSH, 1994; Snook and Ciriello, 1991). This direction yielded several valuable benefits. It identified jobs that could easily be improved through straightforward ergonomics countermeasures, provided good cost justification information related to reducing injuries, and provided positive visibility for the ergonomics program.

Observational task analysis, in combination with a questionnaire and lineside interviews of production associates, were eventually selected as the most appropriate methods for our audits. These methods promoted input from associates and provided an opportunity for objective analysis of manufacturing processes. To perform audits of plant production operations using the selected tools, a second consulting firm was contracted.

For these lineside audit activities we used PDCA, or Plan, Do, Check, Action (Deming, 1986), which is a problem-solving tool regularly used at HAM. We began Planning in detail, meeting with the plants early and often. Once the audits were being implemented (Do), the audit process was reviewed to Check for opportunities to improve it. Action was taken to include these improvements. This PDCA process was repeated at each plant to ensure the best audits possible.

The introduction of our audit methods turned out to be as important as their results. The process of educating associates about the audit laid the foundation for developing an Ergonomics Culture among the production associates. With the establishment of the Ergonomics Culture, the implementation and acceptance of ergonomics countermeasures became a natural progression of the audit itself.

1.4.2 Roles and Responsibilities

The overall responsibility at HAM for completing the ergonomics audits rests with the corporate ergonomics manager who directs corporate ergonomists within the Corporate Health and Safety Department. One of these corporate ergonomists was assigned the responsibility for overseeing the audit at each plant. The safety leader associate at each plant was the point of contact, and he or she was the conduit to resolve issues and gain support for the audit activity within their plant. Each department selected for auditing within the plants had an audit implementation leader associate, and it was this person who scheduled team members and daily audit activity. Much information sharing was required in order to give plant personnel a clear understanding of the audit method, and their after-audit obligation to proceed from audit findings. of ergonomics problems to implementation of countermeasures.

We wanted our associates to gain experience and awareness training during these audits to advance the Ergonomics Culture. Therefore, we arranged the audit teams to include HAM associates who would later be involved in the implementation of countermeasures. By doing this, the associates would not only have some ownership of the audit findings, but would also have a better understanding of how the ergonomics countermeasure ideas related to the findings.

Each audit team consisted of three members: an ergonomics consultant, and from the department, HAM safety and engineering associates. The consultant provided ergonomics expertise and led the audit team. The safety associate provided insight about production associates assigned to the audited process. The engineering team member understood the equipment, tools and process layout. These skills and experiences allowed the team to share knowledge in discussing what they observed, after the lineside observational phase of the audit.

1.4.3 Select Processes to Audit

An automotive manufacturing environment includes many hundreds of processes; some plants have more than a thousand job processes. Therefore, it is unrealistic to attempt a meaningful ergonomics lineside audit of every production process.

A logical approach for selecting processes to be examined in a baseline audit is to identify those processes closely related to the reported ergonomics injuries. After considering all the "leading" and "lagging" indictors that were available, we developed an initial list of targeted processes. Injury data for the last business year were compiled and reviewed for this purpose. We included all ergonomics lost workday injury/illness cases and restricted duty cases.

Two criteria were employed in the selection of processes to audit. These were process injury history and the department's ergonomics priority processes. A threshold number of injuries was established to select processes to audit. Any process exhibiting the ergonomics-related injury of two or more associates within the last business year made the audit list. A review of the department's list of priority processes was compared with the processes selected by injuries. (Each department's priority processes were based not only on injury history, but also on their associate complaints, and/or any previous ergonomics analyses applicable to the process.) Any process not already included by the injury criteria was added to the audit list to ensure thorough audit coverage.

Processes were audited during both day and night shifts of regular manufacturing. Manufacturing operations are essentially the same on day and night shifts. Selection of processes from the audit list was about 75 % from the day shift and 25% from the night shift. This should reflect a realistic representation between the two shifts of production associates.

1.4.4 Assign Audit Teams

Department managers selected an audit leader for their department. Usually, this was the department's safety coordinator associate. These department audit leaders assigned a safety associate and an engineer from each production area in their department to serve on the audit team. Meanwhile, the corporate ergonomist leading that plant's audit scheduled and assigned a third-party consultant to each department to work with their audit team members. These coordinating arrangements required that the department audit leader associate also schedule and arrange for production line associates to be prepared to receive the audit team at their process workplace. During these audits, up to five audit teams worked, with one team in each department. This allowed the team consultant to gain experience with the nuances of the various production processes.

1.4.5 Pre-Lineside Discussions

In preparation for the audit, each team assembled in the audit project room before going to the plant floor. The corporate ergonomist leading that plant's lineside audit provided a list of processes to be audited, together with the specific injury data related directly to each of the processes listed. The department safety associate brought job hazard analysis information and the process's operation standard. The engineering associate provided process and task specific engineering data, like task movement requirements and cycle timing. All this information about the process to be audited was reviewed and discussed with the third-party consultant to assist the consultant in understanding the work to be observed. The front page of the Process Analysis form (see Appendix A) was filled in by the consultant, with identifying and descriptive information specific to the this process and this location . The team was now prepared to start the lineside audit.

1.4.6 Lineside Audits

The observation Process Analysis form facilitated the determination and recording of ergonomics risk factors and contributing task characteristics. Each risk factor was related to a particular body part on the form (Auburn Engineers, 1997). Associate Process Comfort Survey questionnaires (Appendix B) were administered at the time of the audit to the three or four production associates who regularly rotated through that process. The purpose of these questionnaires was to capture the "before" snapshot of a "before and after" record of the work, identifying physical discomfort information from associates working the process being audited. The data to complete these two forms were collected during production time at lineside.

Associates assigned to the audited process were interviewed during lineside observations. They provided first-hand information regarding the nature of the work necessary to perform their manufacturing processes correctly. These associates provided substantial process insight to the audit team. Associates being interviewed were replaced on the production line by alternate associates during the interview and comfort survey process. Associates responded candidly both on the survey and during interviews.

In addition to observations, interviews and questionnaires, the audited process was videotaped from a variety of angles, capturing many cycles of movements and postures. Tools, parts and equipment related to performing the many tasks within that process were also documented.   At least six minutes of videotape documentation was made for each process audited.

1.4.7 Post-Lineside Reviews

Following audit team observations and associate interviews, the Process Analysis form (Appendix A) was completed and the Process Comfort Survey (Appendix B) data were reviewed.   The videotape of the process was reviewed repeatedly. Together with the written notes and observations, all the data were discussed at length among the three-member audit team.   Referring to the analysis form and based on the audit data collected, the team decided by consensus which process activities and postures presented a potential for ergonomics injury. These activities were then recorded as risk factors related to a particular body part and listed on the observational Process Analysis form.

One issue surrounding the determination of ergonomics risk factors by the audit team was the question of consistency among the plant audit team members and the consultants. A standard had to be established and communicated to all audit participants. For our audit purposes, standard examples of work situations were used in observational training exercises by the whole team. The team practiced by observing the same work processes to develop consistency in identifying ergonomics concerns. The corporate ergonomist leading the plant audit facilitated this observational training.

1.4.8 Problem-Solving

The final step in this lineside audit method was for the team to review all of its findings related to the process audited. Working as a brainstorming group, the three-person team applied their problem-solving skills to develop a variety of possible countermeasure ideas. This brainstorming session established a better understanding of the causative factors and how they related to the ergonomics risks identified, while providing insight into the possibility of a solution for each audit finding. These potential countermeasure ideas were provided to each department's production team for feasibility study and prioritizing consideration.

The corporate ergonomist leading the plant audits reviewed the lineside audit results for consistency and to ensure that findings were being properly reported. A clerical support associate entered these audit findings into an internal management software system for the purpose of assigning countermeasure implementation responsibility and for tracking. The Process Analysis forms, Process Comfort Surveys, observation notes, associate interview documents and process videotapes for each process were filed centrally at the plant for later review during countermeasure activity.

1.5 COUNTERMEASURE EVALUATION

HAM has a software program on our internal Intranet called the Safety Task Manager (STM) system, which allows tasks to be assigned to plants and departments. As the name implies, STM provides a method to manage tasks related to safety by assigning task responsibility and tracking task progress. Audit findings are entered into this system and assigned to the responsible department implementation leader for corrective action. The department audit leader is accountable for ensuring that a corrective action plan is developed and submitted within the STM in response to each audit finding. These action plans are reviewed at the corporate ergonomics office, and progress is tracked as the action is implemented. Each ergonomics countermeasure is evaluated by corporate ergonomists for effectiveness, which increases the audit's positive impact.

HAM audit team members are assigned to the countermeasure activity for those production processes that they audit, and they assist the department audit leader with developing action plans. These action plans include the proposed countermeasure and date of scheduled implementation, and they provide a reference for progress reports.

Implemented countermeasures are evaluated on observable reductions in ergonomics risk, improved responses on the Process Comfort Survey questionnaires and/or a measurably lower injury rate. Audit follow-up reviews are conducted at each process to evaluate ergonomics countermeasure effectiveness about a month after each one is implemented. These reviews also determine if any significant risk factors remain to be addressed. The Process Comfort Survey questionnaires are re-administered to get the "after" snapshot, in order to identify any noticeable and consistent improvements.

1.6 AUDIT RESULTS

Many benefits were realized as a result of conducting these audits. Areas for production process improvements were identified. Large gains were made in ergonomics awareness across the corporation. Departments realized significant training and experience advances among the audit team participants. Managers gained a better understanding of the scope and nature of ergonomics problems within their departments. Production staff enjoyed more opportunities for directly contributing new ideas for improving their workplaces. And finally, those responsible to implement the audit countermeasures were encouraged to think "out of the box" and innovative ideas resulted.

1.7 LESSONS LEARNED

Initially, with high production demands, departments could not readily participate in pre-audit training, causing misunderstanding among the audit team members. From this we learned that thorough pre-audit training for audit team members was essential before the audits could be effective. Catch-up training was provided to audit team associates to enhance their understanding of the audit process and improve consistency in analyzing ergonomics concerns and identifying risk factors. Early in the first lineside audit we also recognized that auditing is an ongoing activity that requires continuous fine-tuning. Too much detail in audit findings can confuse the department, resulting in an inability to set appropriate priorities. Too little detail can result in a lack of clear direction and guidance for countermeasure development. The Process Analysis form together with the audit team training resulted in just the right amount of information being captured and reported.

Explaining an audit method, its intent and process to someone does not ensure that they understand or accept it. Candid dialogue should be established among the many levels of management, with feedback opportunities to confirm their support for any audit activity well before it begins. Documenting these activities can avoid misunderstandings later and increases the likelihood of success for both groups.

1.8 CONCLUSION

In auditing you are looking for both beauty and blemishes. Our goal was not only to find problems, but also to help the departments with developing ideas and recommendations for correcting them. An audit's success can best be assessed by how much improvement it brings to the organization audited. The HAM lineside audit method of focusing on solutions rather than just identifying the problems has been successfully completed at two plants to date and is resulting in well-developed Action Plans and positive countermeasure activity.

1.9 REFERENCES

Auburn Engineers, 1997. *Ergonomic Design Guidelines*, (Auburn, AL: Auburn Engineers, Inc.).

Deming, W.E., 1986. *Out of Crisis*, (Cambridge, MA: MIT Center of Advanced Engineering Study).

Drury, C.G., 1998. Auditing ergonomics, In *Ergonomics in Manufacturing*, edited by Karwowski, W. and Salvendy, G., (Norcross, GA: Engineering & Management Press), pp. 397–411.

Kennedy, E.M., MacLeod, D. and Adams W.P.,1992. *Ergonomics Task Analysis*, (Minneapolis, MN: Comprehensive Loss Management, Inc.).

Keyserling, W.M., Stetson, D.S., Silverstein, B.A. and Brouwer, M.L., 1993. A checklist for evaluating ergonomics risk factors associated with upper extremity cumulative trauma disorders, *Ergonomics,* 36(7):807–831.

Kirwan, B. and Ainsworth, L.K. (editors), 1992. Task data collection methods, In *A Guide to Task Analysis*, (London: Taylor & Francis), pp. 41–80.

Kuorinka, I. and Forcier, L. (editors), 1995. *Work related musculoskeletal disorders (WMSDs): a reference book for prevention*, (London: Taylor & Francis).

McAtamney, L. and Corlett, E.N., 1993. RULA: A survey method for the investigation of work-related upper limb disorders, *Applied Ergonomics,* 24(2):91–99.

National Institute for Occupational Safety and Health (NIOSH), 1994. *Applications Manual for the Revised NIOSH Lifting Equation,* DHHS Publication No. 94–110, (Cincinnati, OH: U.S. Department of Health and Human Services, Public Health Service).

Snook, S.H., and Ciriello, V.M., 1991. The design of manual handling tasks: revised tables of maximum acceptable weights and forces, *Ergonomics,* 34(9):1197–1214.

1.10 ACKNOWLEDGEMENTS

The exemplary professionalism and unstinting dedication of the corporate ergonomists in administering these audits is very greatly appreciated. The cooperation of the production facilities of Honda of America Mfg., Inc., was essential in making these audits a success.

APPENDIX A. The recording page of the observational 'Process Analysis' audit form lists body part, risk factor, and the related contributing characteristics. Space is also provided to develop seed countermeasure ideas.

Process Analysis

Process Name:			Responsible Associate:	
Body Part	**Risk Factor**	**Finding No.**	**Contributing Characteristic(s)**	**Seed Countermeasure Ideas**
Hand & Wrist				
Elbow				
Shoulder				
Neck				
Back				
Legs				
Feet				

APPENDIX B. Audit 'Process Comfort Survey' questionnaire illustrates a whole body map and directions for reporting comfort level related to working the process.

PROCESS COMFORT SURVEY

DATE: MONTH DAY YEAR	PLANT:	DEPARTMENT:
TIME: AM☐ SHIFT# PM☐	PROCESS NAME:	
HANDEDNESS LEFT RIGHT	LENGTH OF SERVICE IN THIS PROCESS AREA:	

HOW WOULD YOU DESCRIBE YOUR **OVERALL COMFORT** ON THE PROCESS?
(CHECK APPROPRIATE BOX)

☐1 ☐2 ☐3 ☐4 ☐5 ☐6 ☐7 ☐8 ☐9 ☐10
MOST COMFORTABLE   LEAST COMFORTABLE

WHICH PART(S) OF YOUR BODY IS LEAST COMFORTABLE
AS A RESULT OF THIS PROCESS?

1. Shade Area(s) of Discomfort on the body map

2. Number the shaded areas based on the scale below:

   1 Most Comfortable

   10 Least Comfortable

HAVE YOU SOUGHT TREATMENT FOR YOUR DISCOMFORT?   ☐Y ☐N

IF YES, WHO TREATED YOU?
☐   PLANT MEDICAL
☐   HEALTH PARTNERS
☐   FAMILY DOCTOR
☐   OTHER

Front   Back

PLEASE TELL US MORE ABOUT YOUR EXPERIENCE WITH THIS PROCESS

WHAT MAKES YOUR COMFORT LEVEL BETTER/WORSE?
WHAT DO YOU THINK CAUSES THIS?

DO YOU HAVE ANY SUGGESTIONS ON WAYS TO
IMPROVE YOUR PROCESS?

CHAPTER TWO

Textron Automotive Trim: Initiating a Sustainable Ergonomics Process

Ronald D. Henderson

2.1 INTRODUCTION

Textron Automotive Trim, a division of Textron Automotive Company, has achieved a 77% reduction in Musculoskeletal Injury Rate (MSIR) between 1994 and 1998, with 1994 serving as the baseline value (see Figure 2.1). Initial ergonomics improvement efforts at Textron were based on plant-driven activities and produced varying results. Beginning in 1995, a formal ergonomics process was introduced with consistent analysis methods, training programs, and approaches to problem-solving. This paper describes the critical elements contributing to the success of this initiative as well as lessons learned.

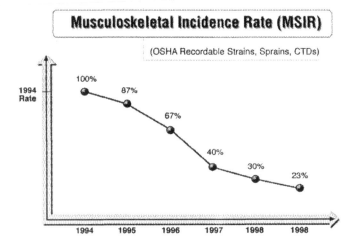

Figure 2.1 Musculoskeletal Incident Rate Reduction at Textron Automotive Trim, expressed as a percentage of the 1994 baseline rate.

2.1.1 Initial Ergonomics Efforts

In 1994, Textron undertook an initiative to better understand the types and causes of injuries and illnesses. The basic assumption was that defining the problem is 80% of the effort of defining the solution. At that time, Musculoskeletal Injuries (MSIs), defined for tracking purposes as Cumulative Trauma Disorders (CTDs), strains and sprains, were a major contributing factor to OSHA recordables (see Figure 2.2). To understand the problem better, the company began tracking a Musculoskeletal Injury Rate (MSIR). In 1994, the rate was calculated to be approximately ten (ten Musculoskeletal Injuries per one hundred full-time employees).

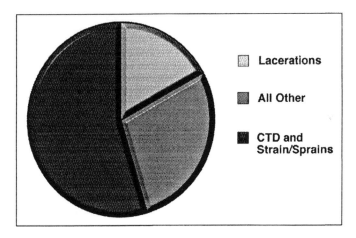

Figure 2.2 Results of the 1994 Injury Trend Analysis.

Plant-based improvement efforts, driven by increased awareness and a Textron philosophy that injuries are preventable and should be controlled, resulted in a 13% reduction in MSIR after one year. Textron Health & Safety staff recognized that an effective problem-solving approach needed to be applied consistently throughout the division to continue achieving improvements in this area.

2.1.2 Formalized Ergonomics Process

An approach was deployed to achieve ergonomics problem-solving skills in key populations and to internalize plant population training delivery through a Train-the-Trainer program. Working with Humantech, a consulting firm in occupational ergonomics, a streamlined training course was customized for engineers, line managers, supervisors, and Health & Safety committee members. Each plant had at least one Health & Safety professional and one engineer trained as a Subject Matter Expert in ergonomics analysis and solution tools. These individuals were also trained to deliver education sessions at their plants: a four-hour "Design for the

Employee" session for engineers, line managers, and supervisors and a one-hour "Ergonomics Awareness" session for operators.

A total of 84 individuals attended one week of ergonomics training delivered on a regional basis (a three-day Applied Ergonomics course plus a two-day Train-the-Trainer program). Participants were equipped with the slides and course materials to deploy the training programs at their facilities. Unfortunately, due to a number of factors, only 20% of the plants deployed the training in the target time frame.

2.1.3 Ergonomics in Design

Even with the limited deployment of plant population training, musculoskeletal disorders were reduced over 50% in the next two years. This was attributed to increased problem-solving skills among the engineering and Health & Safety staffs, as well as continuing top-down support for workplace safety improvements.

The next step was to add an ergonomics review to the capital appropriations process to ensure that new tools and equipment were designed for human performance. By 1998, ergonomics design criteria were integrated into the equipment specification process. Ergonomics design guidelines were published as part of Best Methods Manuals created for common manufacturing processes.

2.2 CRITICAL ELEMENTS OF THE ERGONOMICS PROGRAM

A combination of cultural and technical factors has contributed to the success of Textron Automotive Trim's ergonomics program. Top management has embraced a philosophy that all injuries are preventable and has provided continuing support for Health & Safety initiatives, including the ergonomics program. Three elements stand out as critical to the success of the ergonomics program:
- A two-step approach to ergonomics problem-solving
- Role-targeted training
- Integrating ergonomics into the design of new production processes.

2.2.1 Two-Step Approach to Problem-Solving

Many ergonomics challenges can be addressed with minimum resources and without involvement of engineering staff. With simple ergonomics evaluation tools and years of production experience, operators and supervisors can drive "quick fix" improvements on a daily basis. Frequently, operators will bring an identified ergonomics challenge forward along with their suggested improvement strategy. Working with their supervisor and other experienced operators, they can build upon their improvement idea, coming up with a strategy that is low cost and high impact.

While many ergonomic challenges can be resolved with the talent and resources available to shop floor personnel, some challenges are complex and must involve technical staff such as Health & Safety professionals and engineers. These types of challenges might require capital improvements and must go through the capital appropriations request process. Engineers and Health & Safety professionals have the problem-solving skills and the access to resources to handle complex issues.

This two-step approach to problem-solving is an important element to the on-going success of ergonomics at Textron Automotive Trim. Daily improvements can be made without waiting for the availability of technical staff. At the same time, complex issues can be addressed with the help of technical staff who have the time to devote to ergonomics because they are not fighting daily battles.

2.2.2 Role-Targeted Training

The purpose of ergonomics training is to provide the skills for individuals to fulfill their roles in the company's ergonomics process. Since the ergonomics process at Textron Automotive Trim involves four separate roles, four separate training programs were deployed. These are summarized in Table 2.1.

Table 2.1 Training Courses in the Order Rolled Out at Textron Automotive Trim.

Course	Description
Applied Ergonomics – 3 days Students: Engineers, Health and Safety staff Result: Ergonomics Subject Matter Experts	Prepares technical staff to be Subject Matter Experts in the ergonomics process. Emphasis is on hands-on problem-solving methodologies to recognize, evaluate, and control ergonomics in plant operations. Participants learn to: • Evaluate workstations and equipment for good ergonomic design. • Assess and prioritize ergonomics risk exposure of various jobs and tasks. • Develop high impact ergonomic improvements to reduce risk exposure. • Manage the site ergonomics process.
Train-the-Trainer – 2 days Students: Ergonomics Subject Matter Experts Result: Ergonomics Trainers for plant staff.	Prepares ergonomics Subject Matter Experts to deliver ergonomics training to plant personnel. Participants learn and practice delivering: • Design for the Employee. • Personal Ergonomics Awareness.
Design for the Employee – 4 hours Students: Managers, Supervisors, Engineers, Ergonomics Committee, Health and Safety Committee	Introduces ergonomics as a change management and continuous improvement process. Focus is on: • The job improvement process – emphasizing simple needs assessments and low-cost-high-impact solutions. • The equipment design process – emphasizing simple evaluation tools and easy-to-implement equipment changes.
Personal Awareness Training – 1 hour Students: Operators	Introduces the employee's role in the ergonomics process. Emphasis is on recognizing and correcting ergonomic challenges.

The rollout of the training was sequenced in the order presented in Table 2.1. Subject Matter Experts were trained to solve problems before employees were trained to recognize and report them. This ensured that plant staff would not be overrun with ergonomics challenges that they were not equipped to address.

2.2.3 Integrating Ergonomics into the Design of New Production Processes

New production processes are introduced on a weekly basis throughout Textron Automotive Trim. By addressing ergonomic aspects before workstations and equipment are installed, many challenges can be avoided resulting in a low-cost or even cost-free solution. Three important elements were deployed to ensure that ergonomics is integrated into the design process:

- Ergonomic reviews of workstations and equipment have been added to the Environmental, Health & Safety checklist that was already required as part of the Capital Appropriations process. Each new piece of equipment is now evaluated according to ergonomic design and build guidelines, and identified deficiencies are addressed before final sign-off on the design.
- Ergonomic designs have been added to the Best Methods Manuals that had been developed for common production processes. These "templates" now provide engineers and manufacturing personnel with a starting point of a good ergonomic design, so that they do not have to rely on existing equipment set-ups, which might or might not represent good ergonomics.
- Training has been provided for engineers and product development teams in ergonomics and available resources. This provided design staff with the skills to properly apply the information that had been organized for their benefit.

2.3 LESSONS LEARNED

The deployment of a formalized ergonomics process at Textron Automotive Trim has not been a one-shot approach. Lessons learned over the past six years have resulted in the discovery of new leverage points, and their exploitation for continual improvement of the approach. Three important considerations that can be applied to every company's ergonomics initiative are described below.

2.3.1 Cross-Functional Ergonomics Process with Clearly-Defined Roles

Cross-functional problem-solving has proven to be effective for initiatives such as Total Quality Management and Lean Manufacturing. Within an ergonomics initiative, cross-functional problem-solving can begin with simple awareness training programs for supervisors, engineers, and operators that emphasize real life challenges and solutions. Problem-solving can be accelerated through the introduction of usable tools targeted to roles in the ergonomics process. These roles can be broken down into Recognition, Evaluation, Control, and Anticipation:

- Recognition – Simple methods to identify and communicate ergonomic challenges for operators and supervisors.
- Evaluation – An ergonomics risk assessment that evaluates the level of risk and contributing factors for Health & Safety professionals.
- Control – Problem-solving methods that lead to cost-effective solution strategies.
- Anticipation – Ergonomic design guidelines for engineers and design teams.

2.3.2 Effective Training

Training is an important component of an effective ergonomics process. For maximum effectiveness, training should be:

- Performance-driven – Training should provide the skills for participants to fulfill their role in the ergonomics process.
- Relevant – Site-specific case studies and hands-on data collection ensure that the training program keeps the attention of the participants and provides them with the opportunity to confirm that they can deploy the ergonomics process.
- Delivered by effective trainers – Trainers should be comfortable with both training delivery and the technical content. This can be achieved through Train-the-Trainer sessions and by selecting experienced trainers to deliver the education sessions.

2.3.3 Ergonomic Design Criteria

Resolving ergonomic challenges in existing operations is necessary, exciting, and can make a real impact on the health, safety and performance of operations. However, in rapidly-changing manufacturing companies such as Textron Automotive, solving existing problems is only half the battle. A parallel effort is needed in the design and selection of workstations, tools, and equipment.

Experience has shown that it is much more cost effective to "build in" good ergonomic design, rather than to "bolt on" solutions to address poor ergonomic design. As equipment moves through the stages of concept, design, prototype, and production tool, the costs of changes go up, and the constraints to change increase rapidly. Integrating ergonomics criteria early in the design stage can be a very low-cost, very high-impact approach to improving the ergonomics of the work place.

Ergonomics design criteria can be shared in the form of Design and Build Guidelines. This is a simple approach to getting basic human performance criteria into the hands of those individuals who are specifying new tools, workstations, and equipment so that they can communicate the company's needs to suppliers. Figure 2.3 illustrates one approach used by Textron Automotive Trim, a design checklist.

TEXTRON Automotive Trim
Ergonomics Design Criteria

Checklist for Standing Workstation Criteria

Standing workstation criteria	Specification	Actual measurement	Within specification		Action(s) required
			Yes	No	
Work distance					
One Hand (Figure 3)	4 to 18 inches from table edge				
Two Hands (Figure 4)	4 to 16 inches from table edge				
Worksurface dimensions (Figure 5)					
A. Height	Heavy Assembly: 29 to 39 inches				
	Light Assembly: 35 to 45 inches				
	Precision: 40 to 50 inches				
B. Angle	Adjustable from -6° to 30°				
	(-) = away from operator				
	(+) = towards operator				
C. Rounded Edges					
D. Footrail Height	6 inches				
E. Knee Clearance	Minimum of 4 inches				
F. Foot Clearance	4 inches high 4 inches deep				
Anti-Fatigue Matting	At least 1/2 inch thick				
Top of Monitor	56 to 70 inches from floor				
Head Clearance	Minimum of 80 inches				

©Humantech

Figure 2.3 Ergonomics Design Checklist from Textron Automotive Trim.

2.4 THE NEXT STEP FOR ERGONOMICS AT TEXTRON AUTOMOTIVE TRIM

Initially, the value of ergonomics was seen as a purely Health & Safety benefit; management wanted incident rates and workers compensation costs to come down, and MSIs were found to be a major contributor to these metrics. The ergonomics initiative has been a very important part of Textron Automotive Trim's success in driving down total injuries (illustrated in Figure 2.4).

As ergonomics has become more of a "way of doing business," integration with operational initiatives such as Lean Manufacturing has driven continuous improvement.

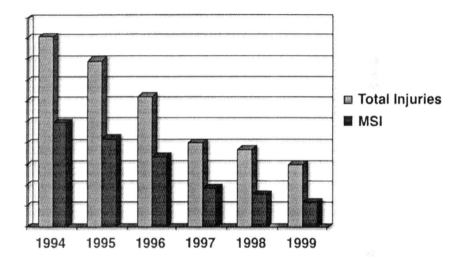

Figure 2.4 Recordable incidents continue to decline from 1994 levels.

2.4.1 Lean Manufacturing

Lean Manufacturing, a process improvement approach with the goal of lowering costs by eliminating waste, has become a high level initiative within Textron Automotive Trim. Work in process inventory is being driven down, cycle times are being compressed, and changeovers are quicker than ever.

Properly deployed, ergonomics benefits Lean Manufacturing in a measurable way. As the extreme postures and forceful motions are designed out of jobs, cycle times can be reduced without a negative impact on the operator. Ergonomics is now seen as an effective way to optimize motions while increasing the health and safety and morale of the work force.

The language of Lean Manufacturing is time. Consequently, Humantech's Standard Time Efficiency Process (STEP©) is being added as a methodology delivered during the Applied Ergonomics training course. The STEP allows Subject Matter Experts to quantify the time-savings from ergonomics improvements, providing a basis for cost-benefit analysis to drive more ergonomics improvements for work areas where injuries have not yet occurred.

2.4.2 Deployment of Ergonomics with Textron's Lean Initiative

A new easy-to-use reference booklet has been developed for use during the lean events that are occurring across Textron Automotive. This 16-page booklet emphasizes the basic ergonomics principles that affect motion, and provides specific guidelines for workstation setup and tool selection. It fits ergonomics criteria into the standard lean approach of the visual workplace and provides the STEP methodology for calculating time-savings from ergonomic improvements.

Ergonomics is an important Health & Safety continuous improvement tool. In addition, it is becoming an important part of the Lean Manufacturing initiative. By putting guidelines for safe working height and reach distances side-by-side with guidelines for efficient work zones, engineers and other personnel can fully understand the complimentary benefits of ergonomics.

Over the past six years the ergonomics process at Textron has matured and adapted to changes in manufacturing strategies. The future of the ergonomics process will focus on the integration of ergonomics into Lean Manufacturing, and into mistake-proofing and error-proofing.

CHAPTER THREE

Ergonomics Before Equipment Receipt

Keith M. White, Ph.D., CPE and Paul Schwab

3.1 ABSTRACT

A wafer fabrication facility was recently upgraded from 150mm (6-inch) to 200mm (8-inch) wafer production, a project that involved the purchase of a considerable amount of capital wafer processing equipment. Although industry ergonomics guidelines existed, and the manufacturer's application of these guidelines was a corporate requirement for the purchase of new equipment, there was not an effective tracking system in place to ensure that all new equipment met these guidelines. The ESHEER (Environmental, Safety, Health and Ergonomics Equipment Report) process was developed as a means to track and document safety and ergonomics issues associated with new equipment. This process has been an effective means of ensuring that equipment suppliers conform to established industry guidelines.

3.2 SEMICONDUCTOR MANUFACTURING PROCESS

Integrated circuits have found their way into many of today's products, from computers to cell phones to electric motor controllers. The heart of these integrated circuits is a small semiconductor "chip" no bigger than a few square millimeters. Thousands of these chips are produced on a "wafer," a relatively thin slice of very pure silicon. The device is called a semiconductor because silicon, under the right conditions, can conduct electrical currents. During the manufacturing process, various means are employed to improve the semi-conductor properties of the silicon.

Wafers have grown in diameter as the technology to produce the wafers has improved. Some of the earliest wafers were about 25mm (1 inch) in diameter while the latest endeavor is a 300mm (12 inch) wafer, about the size of a small pizza. There are many chips built on a silicon wafer, so the larger the wafer, the greater the surface area and the greater the number of chips that can be produced on the wafer.

The wafers are processed on sophisticated equipment called process tools or "tools" for short. Figure 3.1 is an example of such a tool. These tools can cost several million dollars each and a typical wafer fabrication facility (wafer fab) can house several hundred tools totaling billions of dollars.

Figure 3.1 Semiconductor process tool (Applied Materials 5200).

3.3 PROJECT BACKGROUND

A wafer fabrication facility was recently upgraded from 150 mm (6 inch) to 200 mm (8 inch) wafer production, a project that involved the purchase of a considerable amount of capital wafer processing equipment. This project was known as the DIAMeter TECHnology or DIAMTECH project. This project entailed getting about 70 new types of semiconductor processing tools from about 31 tool manufacturers or vendors. The acquisition of new tools gave us an opportunity to ensure that the design of the tools followed sound ergonomics principals and would accommodate the majority of the user population.

An ergonomist was assigned to the DIAMTECH project whose primary responsibility was to ensure the new 200mm wafer processing equipment to be received for this project was compliant with the ergonomic principles of the semiconductor industry guidelines. The goal was to avoid the problems experienced with previous generations of semiconductor equipment used in the legacy 150mm wafer fab, equipment which had been designed with little consideration for the users (see Figure 3.2).

Figure 3.2 Equipment should be designed with consideration for human abilities.

3.4 SEMICONDUCTOR INDUSTRY ERGONOMICS GUIDELINES

Semiconductor production is very capital-intensive and requires sophisticated technology from multiple suppliers so the industry has benefited from global standards created by Semiconductor Equipment and Materials International (SEMI®), a global trade association that represents the semi-conductor and flat panel display equipment and materials industries. SEMI was founded in 1970 in the United States and has evolved into a worldwide organization with representation around the globe. SEMI guidelines have helped to standardize materials, equipment and services in the semiconductor industry, thereby reducing the overall cost and increasing the reliability of products and services.

SEMI safety and ergonomics guidelines foster standardization between the suppliers and users of semiconductor manufacturing equipment, supplies and support facilities. These guidelines were developed through the SEMI International Standards Program as a consensus within the industry. In 1993, general ergonomics safety guidelines were specified in *SEMI S2, Safety Guidelines for semiconductor Manufacturing Equipment.* In 1995, more prescriptive recommendations were specified in the *SEMI S8-95 Safety Guidelines for Ergonomics Engineering of semiconductor Manufacturing Equipment document.* A detailed checklist was added to the September 1999 version (see Figure 3.3 for an excerpt from this document).

The SEMI guidelines recommend that an independent third party evaluator evaluate each tool. For ergonomics, the third party evaluator uses the *SEMI S8 Safety Guidelines for Ergonomics Engineering of semiconductor Manufacturing Equipment.* Included in the S8 guideline is an appendix called the Supplier

Section	Indicator	Acceptance Criteria Metric units (US Customary units)	Actual
3	**Wafer cassette loading**		
3.1	Wafer cassette loading should not require greater than 10 degrees cassette rotation in any axis.	less than 10 degrees rotation in any axis.	
3.2	Load port height, vertical distance from standing surface (150 - 200mm wafers).	maximum 960 mm (38 in) minimum 890 mm (35 in)	
3.3	Maximum lip height in front of cassette load port over which cassette must be lifted (150 - 200mm wafer cassettes only).	maximum 30 mm (1.2 in)	
3.4	Reach distance from the leading edge of the tool or obstruction to the coupling point(s) on rotation device or the product grasp point.	maximum 330 mm (13 in)	
3.5	Minimum hand clearance on either side of the cassette, measured from the side of the cassette to the nearest adjacent object.	minimum 76 mm (3.0 in)	

Figure 3.3 Sample of the Supplier Ergonomic Success Criteria (SESC) checklist excerpted from Appendix 1 of SEMI S8-0999.

Used with permission from Semiconductor Equipment Materials International. This document and other SEMI International Standards are available for purchase from SEMI at www.semi.org.

Ergonomic Success Criteria (SESC) checklist. An independent third party evaluator will typically use this checklist to collect much of their field data, take it back to their office to interpret the findings and then publish the findings in a *SEMI S2-93A Product Safety Assessment*. During this process, there is considerable dialogue between the third party evaluator and tool manufacturer to clarify and, hopefully, resolve any discrepancies.

The goals of the SEMI-S8 guidelines are to insure that the design of equipment meets the following criteria:

- Accommodates a diverse user base (90% of the world's working population).
- Allows users to work in a comfortable, neutral posture.
- Accommodates user visual limits.
- Facilitates maintainability.
- Avoids user overexertion due to manual handling tasks (lifting, pushing, pulling, etc.).
- Provides user-friendly interfaces.

As a whole, the Semiconductor industry has embraced SEMI guidelines for equipment manufacturers to use in designing, building and assessing their tools. Fortunately, because of demands from their other customers, most of the tool suppliers were already familiar with the semiconductor industry's ergonomics standards so they were not surprised by the importance of ergonomics requirements for equipment specified for the DIAMTECH project.

3.5 TOOL PROCUREMENT PROCESS

An example of the tool acquisition process for the DIAMTECH project follows. Unfortunately, for most tool acquisition procedures at the beginning of the DIAMTECH project, this process was not followed exactly (see Figure 3.4). To better track this acquisition process while the project progressed, a tracking system was developed called the ESHEER, an acronym for: Environmental Safety Health Ergonomic and Equipment Report, an internal document that is explained further in this paper.

The ESHEER form is an electronic document created in a word processor format. Individual documents are created for each tool. The electronic documents are stored on a server and are accessible to all of the Environmental Safety and Health Professionals in the company. The advantage of this format is that it is simple to use and text and images from third-party reports or other sources can easily be inserted. The other advantage to this system is that the files can be updated immediately during tool negotiation meetings or phone conversations. Copies of the reports are made available to the tool suppliers so that they know what actions to take.

Descriptions of the key steps in the tool acquisition process follow.

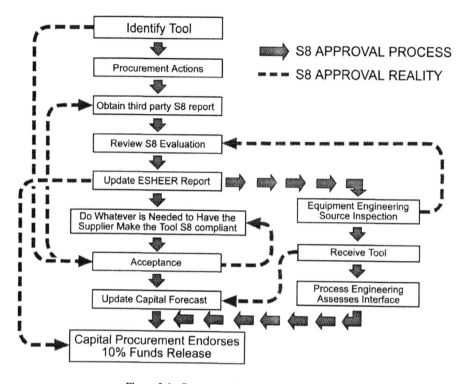

Figure 3.4   Process tool procurement process.

Identify tool: Many people are involved with identifying tools for a specific process need. Ideally, there would be more than one tool supplier candidate for the process to allow for alternate suppliers.

Procurement actions: This includes sending the tool supplier documents detailing wafer fab requirements and soliciting comments from the supplier as to their ability to meet each requirement, setting up meetings for the appropriate wafer fab personnel to discuss technical concerns their about the tool(s) and conducting commercial negotiations.

Obtain third party evaluator S8 report: The tool manufacturer supplies the third party evaluation report to the appropriate safety and ergonomics representatives.

Review S8:   Items to be addressed are listed on the ESHEER form. These items are obtained from the third party reports and from the experiences of other fabs with similar tools.

Update ESHEER form: The ESHEER form is updated by the Ergonomist or Safety Professional.

Do whatever is needed to have the supplier make the tool S8 compliant:
Texas Instruments is committed to optimizing the ergonomics of tools and we are willing to work with the tool manufacturer to ensure the best user interface for our employees. Ideally, the needed change(s) would occur before the tool is shipped, however, this is not always possible. The need to maintain tool delivery dates in order to maintain the project schedule to meet business objectives is critical. In these cases, the tool user (manufacturing) is required to make the tool available for a retrofit by the supplier for no more than two weeks when the ergonomic correction is available. Before committing to a retrofit, the manufacturer is required to ensure their correction can be accomplished in a two-week time period for each tool.

Acceptance: Once all items are resolved, the third party evaluator re-evaluates the tool and, if passed, typically indicates that the tool is in compliance with the SEMI S8-95 guideline. In many cases, the third party evaluator evaluating the tool issues a "Certification of Compliance." At this point, acceptance of the tool is achieved.

Update capital forecast: Upon acceptance, the capital forecast sheet, which tracks procurement, delivery and accounting information, is updated.

Capital procurement: As part of the purchase contract, 10% of the original purchase price of the tool is withheld from the supplier until the tool meets SEMI S2 and S8 guidelines. If any outstanding items need to be resolved, the third party evaluator re-evaluates the tool and, if it passes, the remaining 10% of the tool purchase price is released. This withholding insures that the tool supplier makes any and all changes that were promised.

Equipment engineering and source inspection: In many instances, the Equipment Engineer (EE) responsible for the tool has an opportunity to inspect the tool before shipment to the wafer fab. The EEs are given a one page ergonomic checklist to use when inspecting the tools. The EEs are also provided with simple instructions: "If anything about the tool seems awkward, it probably is, so please let the project ergonomist know."

Receive Tool: The Process Engineer qualifies the tool and ensures that it can correctly process wafers. This is accomplished by processing many wafers through the tool and measuring the results on the wafer. In essence, the process engineers are the first tool users. As such, they were also asked to identify anything awkward with the user interface of the tool.

3.6 EXAMPLES OF EQUIPMENT MODIFICATIONS

Some examples of ways potential ergonomics problems were resolved for the DIAMTECH project follow.

1. The loadport design of a wafer polishing tool resulted in the user gripping the product in an awkward manner and using a stooping posture to load and unload material submerged in a water tank. The process tool supplier was able to modify the loadport user interface to eliminate awkward postures used to load and unload material. With the new interface, instead of reaching into a tank to access the material, the user now opens a door that presents the material at elbow height. As an added benefit, this equipment modification has a predicted increase in product throughput. Because of a tight schedule, the supplier was unable to make the modifications to the first two production tools delivered. A short-term retrofit that had been proven at another wafer fab was added to the load stations of the first two production tools until they could be modified. A sliding shelf allowed users to place a carrying tray near the unloading station so that they could change hand positions between unloading and carrying material, thereby reducing awkward grips.

2. A metrology (measuring) tool had a loadport lift-over height 114 cm (45 inches) above the floor. The SEMI S8-95 guideline recommends a loadport height of 89 cm (35 inches) to 96 cm (38 inches) above the floor. To resolve this problem, the tool manufacturer cut away an obstruction to expose the loadport and replaced the cut-away section with a hinged door so that there would be no exposed moving parts during machine operation. This modification reduced the lift-over to an acceptable height. To load the tool, the operator places the wafer cassette on a staging table, releases the hinged access doors, places the material on the loadport and then closes the access doors.

3. The loadport of another tool could not be redesigned to lower it 10 cm (4 inches) to meet S8 guidelines. The solution was to lower the section of the floor where this tool was to be placed. This tool was very large and required support modifications to the floor where it was to be placed anyway. In a clean room, the floor is perforated to allow the air to flow downward making the air cleaner than the cleanest hospital surgery room. To obtain such a particle reduction, the flooring, which has holes (or waffles) in it, is raised above the foundation. Since this tool required support modifications to the floor, the tool installers simply made the support platform 5 cm (2 inches) lower. This "lowered" the tool below the walking surface so that the loadport was within an acceptable height range.

3.7 ENVIRONMENTAL SAFETY HEALTH ERGONOMIC AND EQUIPMENT REPORT (ESHEER) PROCESS

The ESHEER document is used to record and track the tool acquisition process. The ESHEER document is intended to be a communication tool and serves to record actions between the tool supplier and customer. Beyond tool acquisition, it is a centralized document that each wafer fab will eventually be able to access to review and update. Elements of this document are detailed below with some specific examples from an ESHEER for an actual tool.

3.7.1 ESHEER Report Elements

I.	Equipment Information
II.	Third Party Evaluator Information
III.	Safety, Environmental, Health and Ergonomics Evaluation Information
IV.	Ergonomic Evaluation
V.	Environmental, Safety and Health Contacts
VI.	Supplier/Vendor Contact
VII.	Texas Instruments Contacts
VIII.	Tool Layout
IX.	Equipment Safety-Related Injury/Illness
X.	Equipment Ergonomic-Related Injury/Illness
XI.	Manufacturing Operating Procedure
XII.	Chronology of Events
XIII.	Human-Related Scrap Events
XIV.	Lift Activities

I.   Equipment Information

The equipment information simply identifies the tool manufacturer, model number and the type of equipment. Tool serial number and other identifiers are recorded.

II.   Third Party Evaluator Information

The company that performed the independent third party evaluation to SEMI guidelines, the date of the report and the report number(s) are recorded. This section proved to be highly valuable since many versions of the same tool existed.

III.   Safety, Environmental, Health Evaluation Information

Typically, the third party evaluator report combined safety, environmental, health and ergonomic information. Since ergonomic considerations can be unique and the person who addresses safety concerns is not necessarily the same person who addresses ergonomic concerns, the safety and ergonomic information was separated into distinct sections on the ESHEER (see section 3.4, Ergonomic Evaluation below). Tool ergonomics, safety, environmental and health documentation included the following:

1. Ergonomics and safety concerns identified by the third-party evaluator.

2. Abatement/Acceptance descriptions of what must occur for the tool manufacturer to resolve the concern. This is typically outlined by the third party evaluator. After the tool manufacturer resolved the concerns the corrective action was recorded.

3. Date the correction was verified by the ergonomist or safety professional.

Figure 3.5 shows an example of an environmental/safety concern and corrective actions taken.

3rd Party Identified concern	Abatement/Acceptance	Date
1. 20.3.1 Information Required: Exhaust effluent should be characterized of ADH. (ADH effluent is composed of Ammonia, TMS, HMDS, HMDSO)	**To be performed 300mm evaluation.** (Supplier) must supply a complete analysis of exhaust effluent for one cycle of (all) processes. (all chemicals) (Supplier) standard recipes can be used (Due Friday 1/28/00). Process exhaust and waste effluents were characterized, and mass balance sheet regarding HMDS, butyl acetate, MAK and NMD3 was documented for environmental evaluation. (Completed 12/14/97)	2/8/00

Figure 3.5 Safety, Environmental and Health Evaluation Information.

IV.   Ergonomic Evaluation

Two sections are provided in the ergonomics evaluation. The first section addresses third party evaluator identified concerns as described above and the other addresses any concern identified by the user during source inspection, installation, qualification or use.

Figure 3.6 documents a user concern that was identified during a source inspection and a means to help resolve the problem. Although the reach distance to the product met S8-95 guidelines, there was an overhead obstruction, which obscured the line of sight for taller users. Figures 3.7A and B illustrate the concern and Figure 3.7C shows a recommendation made to the supplier, which was implemented on future tools.

Figure 3.8 documents a bottle change-out activity, which requires the user kneel on the floor. A portable kneeling pad was provided and made accessible on a bottle delivery cart (Figures 3.9 A and B).

V.   Environmental, Safety, Health and Ergonomics Contacts

Environmental, Safety, Health and Ergonomics contact information is provided to ensure that everyone knows who the appropriate people are and how to contact them.

TI Party Identified concern	Abatement/Acceptance	Date
Per Ryan Priebe (EE) source inspection (16 SEP 99), Front lip limits loadport view of a tall operator.	Supplier is negotiating modification of top panel as paid option since the concern is not related to S8 non-compliance (see figures 7A-C).	

Figure 3.6 Ergonomics evaluation: loadport access.

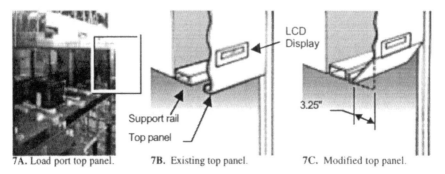

7A. Load port top panel.   **7B.** Existing top panel.   7C. Modified top panel.

Figures 3.7A-C Recommended load port header modifications.

TI Party Identified concern	Abatement/Acceptance	Date
Bottle change on lower shelf requires operator to kneel on waffle floor.	K.White to provide drip tray that includes a means to catch resist and reduces localized pressure on knee. Evaluated by QRA and approved for DM4 use.	

Figure 3.8 Ergonomics evaluation: bottle change-out procedure.

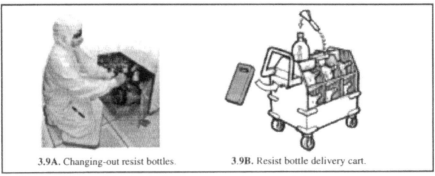

3.9A. Changing-out resist bottles.   3.9B. Resist bottle delivery cart.

Figure 3.9 Kneeling pad recommendation.

VI. Supplier/Vendor Contacts

Originally, this was a small section to identify key support people for the tool manufacture such as site contacts, regional contacts, corporate safety and ergonomic representatives. The number of contacts increased as the DIAMTECH project progressed and we started to work with various tool supplier support groups. It was critical to include the person's job title, phone number, mobile number or pager, and E-mail address.

VII. Texas Instruments Contacts

Contacts within the company are provided to promote discussions between various Texas Instruments wafer fabs. Representatives include: Equipment Engineers, Manufacturing (typically a production area Superintendent who is two levels of management above a tool operator), Process Engineer and Procurement Specialist.

VIII. Tool Layout

This section is used to document pictures or diagrams of specific ergonomic concerns. Typically, at least the machine operator interface zone is displayed.

IX. Equipment Safety-Related Injury/Illness

Within this section, any safety-related event associated with the use of this tool is documented. Particular attention is focused on the event description and any corrective actions.

X. Equipment Ergonomics-Related Injury/Illness

Similar to safety, any ergonomics-related injuries or illnesses are documented.

XI. Manufacturing Operating Procedure

Since the tools were not ready for wafer processing, tasks were simulated and videotaped. The videotapes were reviewed by an independent occupation therapist (OT) who had considerable cleanroom experience. The OT reviewed the process flow and videotape to identify the upper extremity tissues used to perform the task. The OT then considered the frequency of the task and assessed the relative risk that the task would contribute to a musculoskeletal disorder. Each assessed task to date has been considered a <u>low</u> risk, which helps to validate the SEMI S8 guidelines. An example of an assessment of an operating procedure is detailed in Figure 3.10.

Activity	Recommendations
Operate tools with SMIF Pod 1. Assume frequency of operations at one per 5 min. 2. Load/unload rack and place from/to 35-38"surface 3. Place from 35-36" surface to a 35-38" surface	Limit extreme reaching where the elbow is extended with a load in the hands. Employees need to use good work technique to keep load close to body. Consider alternating between loading to the right side of the body and the left side of the body.

OPERATION: Load/unload rack and place from/to 35-38 inch surface
NOTE: Manufacturing protocol dictates a person to use shelf heights with which they are comfortable: i.e. they should get help if they cannot comfortably reach top shelf.
NOTE: Virtually all of our cart, table, and tool load port heights are in a range from 35' to 38'.
NOTE: Shelf heights on the rack will be 19", 34", and 48".

Motion	Tissue(s)	Potential MSD	Risk
Unload rack to cart/table/tool		**If high rating:** Trigger thumb Trigger fingers CTS Flexor/extensor tendinitis	Low
Grasp SMIF Pod from shelf.	Thumb and finger flexors; thumb adductor		Low
Lift SMIF Pod slightly and bring close to body at comfortable carry position.	Elbow flexors/extensors; finger flexors		Low
Turn with feet or move laterally.			Low
Carry to destination. Note: Distances vary considerably	Static contraction of finger flexors/extensors; Elbow flexors/extensors		Low
Place SMIF onto cart/table/tool	Elbow extensors/flexors		Low

Figure 3.10   Manufacturing operating procedures.

XII.   Chronology of Events

This section records key conversations or messages. This further ensures that all wafer fabs with a common tool set can know the most recent dialogues addressing the concerns (see Figure 3.11).

From: (Supplier Representative), [mailto:representative@supplier.com]
Sent: Monday, February 07, 2000 11:48 AM
To: White, Keith
Subject: RE: (Tool) Status

Dear Keith,

I have received the (Tool) 300 mm adhesion exhaust data. I incorrectly reported below that it was not available.

This data is representative for concentrations that will be found on both 200 and 300 mm toolsets. I am forwarding this document to (TI procurement) by fax today.

Please let me know if the Safety group has any questions.

Sincerely,
(Supplier Representative)

Figure 3.11 Memorandum example from "Chronology of Events" section of ESHEER.

XIII. Human-Related Scrap Events

The profit from one wafer varies considerably but, in general, one wafer can be equated to the price of a mid-size car. Breaking a wafer or making it unusable is like taking a new car to a junkyard and having it crushed. Like a car, a scrapped wafer is recycled but the value of the material is much lower than the finished product. Because of the high cost of the end product, it is essential to learn from the events that cause wafers to be scrapped (see Figure 3.12). The focus on scrap caused by human error relies more on human factors engineering principles than ergonomics because the errors are usually cognitive.

DATE	9/10/98
AREA	Photolithography
WAFERS	17
REPORT-ID	DM5-QC-9810-065
TOOL	Coater/Developer CD200
EVENT	Verbal miscommunication between equipment engineering technician and the manufacturing specialist: "wafer got stuck in tool. The equipment engineer did not provide sufficient information for equipment operator to accomplish their portion of the recovery"

Figure 3.12 Human-related scrap.

XIV. Lift Activities

Most third party evaluations, especially those evaluations performed within the last few years, include lift-related information, usually for tool repair activities or equipment installation. Most of the lifts are assessed using the NIOSH lift calculation, psychophysical capacity data or other analysis tool. A matrix of the results of third party lift analyses for various machine components is shown in Figure 3.13. The columns are described in the bottom section of the form.

Item	Lbs.	R NR	RWL	2PAC RWL	Lift	2PAC MA Access	V. Doc Pg. #	Label Y/N	EE Check-List	Comment
Process Block, wafer carrier w/ 25 wafers	5.6	R	18.6 *PCD		1	Y				
Ext. Chemical Cabinet- resist, 1 gal. Bottle/ upper stage	11.7	R	19.5* PCD		1	Y				
Ext. Chemical Cabinet- resist, 1 gal. Bottle/ lower stage	11.7	R	21.5* PCD		1	Y				
16ch Multi Thermo-Controller, side panel enclosure	22.2*	R	21.8	39.24	2	?*				R-every 3-6 months *-Engineer estimates the weight to be <=10lbs

PCD-Psychophysical Capacity Data

Page 2: Lift activity descriptions	

Item: List the item to be lifted
lbs: List item weight in pounds
R/NR: Is the task Routine or Non-Routine?
RWL: Recommended Weight Limit (lbs.)
2PAC RWL: Recommended safe lift for 2 people (lbs.) (2xRWL-10%)
Lift: Type of lift being performed
               1=1 Person
               2=2 Person
               MA= Mechanical Assist documented by vendor
TIMA=   Mechanical Assist by TI to equate the 2 Persons need
V.Doc Pg #: Location of warning in Vendor Manual.
               i.e. 1:3-26, 5:4-12
Footnote: vendor document referencing appropriate lift page
               Page: reflect page number in the document
Label: Is a label on the item? Yes/No
               Is the label appropriate? Yes/No
Access: Is there enough space to perform the operation as given by the Lift Type?
EE Checklist: Is Lift Type 2, MA, or TIMA in the EE checklist?
Comment: Especially comment on type and quantity of MA means and devices.
	Items above reference **Lift Analysis in SESC**

Figure 3.13 Lift activities.

3.8 SUMMARY

Two key elements ensured success of the tool acquisition process: the SEMI accepted guidelines and management's requirement that tools must meet these guidelines. For tool acquisitions, the ESHEER is a communication tool that simply consolidates information to keep the process on track. Long-term benefits rely on obtaining injury/illness and wafer scrap information and sharing the prevention of these events with other wafer fabs.

While the ESHEER tracking process has only been in use for 7 months, we have enjoyed overall success.

1. Several tool manufactures have used the ESHEER forms and other summary information to present their status to their Division and Corporate management.

2. Tool negotiations have begun to outfit a new 300mm wafer fab. Most of the tool suppliers who have been through the 150mm to 200mm conversion requested that we use the ESHEER process for the new facility.

3. During some of the tool negotiations, a few of the local supplier representatives actually presented ESHEER information to their staff. One person concluded his staff briefing by saying this process was a win-win situation and another told his division representatives that this was the right approach for everyone.

3.9 ACKNOWLEDGEMENTS

Several tool manufacturers provided valuable input and professionally challenged the ESHEER approach so that it could evolve into a viable and efficient process. Manufacturers that provided considerable input include Tokyo Electron America, and Eaton Corporation–Semiconductor Equipment Division. We would especially like to thank Applied Materials and their Dallas region representatives for their support. Not only did Applied Materials embrace the ESHEER process and aggressively close all but a few third party concerns, they also demonstrated that the process can be beneficial for both the tool manufacturer and the customer. Recognition is also extended to HealthSouth who assessed the simulated activities to predict if a problem task existed. Finally, these acknowledgements represent the teamwork needed to ensure the ergonomic soundness of tools and operator actions, even before the equipment is received.

SEMI S8 and other SEMI International Standards are available for purchase from SEMI at www.semi.org. NOTE: SEMI International Standards are subject to revision at any time. For information on or to obtain the most up-to-date version of these standards, please visit SEMI at www.semi.org.

3.10 REFERENCES

Semiconductor Equipment and Materials International, 1999. SEMI S8-0999 Safety Guidelines for Ergonomics Engineering of Semiconductor Manufacturing Equipment. Mountain View, California.

Waters, T.R., et al., 1994. Applications Manual for the Revised NIOSH Lifting Equation. Cincinnati, OH: US Department of Health and Human Services.

Mital, A., Nicholson, A.S., and Ayoub, M.M., 1993. A Guide to Manual Materials Handling. Taylor and Francis: London.

CHAPTER FOUR

Sharing Ergonomics Solutions in a "Virtual Factory" System at Intel Corporation

Theresa House and Angela Harrison

4.1 INTRODUCTION

"In a multinational corporation, how do you share ergonomic learning and drive solutions to factories located in different countries?" Any company that has multiple worksites has faced this dilemma at some point in their ergonomics program. At Intel, if you go to a factory in New Mexico, you will see the same ergonomic solutions implemented as in California, Ireland, or Israel Intel factories.

How does Intel ensure that solutions are not only shared, but also implemented at different sites? The system Intel has created is referred to internally as a "Virtual Factory." This system assumes that all factories have some commonality, enough to share solutions and implementations. Intel has driven this Virtual Factory philosophy into all aspects of our business: engineering, safety, and manufacturing. Ergonomic solutions are developed, validated, and implemented across all manufacturing facilities using this system. The result is that a solution can be "copy exactly" across multiple sites, thereby eliminating re-inventing the wheel.

This chapter will outline how ergonomics can embrace the "Virtual Factory" "copy exactly" mentality to realize a cost-effective, highly productive, and time efficient solution.

4.2 SYSTEMS NEEDED

Intel realized that the first thing they needed in order to share ergonomic solutions was a method to discuss ergonomics issues across the sites. A key decision was made to implement a Virtual Factory approach to resolving ergonomics issues. The basic elements included: the formation of a Virtual Factory ergonomics team; a defined communication process; a defined decision-making process; robust solution development; and methodologies for solution testing and implementation.

The system is structured to work in the following way. A team of ergonomics-focused employees from various Intel sites agrees upon critical ergonomics improvements needed at multiple factories. These improvements are documented on a team roadmap and communicated to management and key stakeholders. The ergonomics team members then partner with Intel engineers to develop, test, and implement solutions across multiple sites. The following sections depict the details of each of the critical elements of a Virtual Factory ergonomics team approach.

4.2.1 Virtual Factory Ergonomics Team

A Virtual Factory (VF) ergonomics team was formed with a defined charter and membership, as well as defined strategic objectives and deliverables. The team charter was "To resolve high priority ergonomic risk factors across multiple manufacturing sites." The membership consisted of one person from each site whose job description included resolving ergonomic issues for his or her site. In addition, there was a management level team coach whose role was to ensure the VF ergonomics team was aligned to Intel business objectives.

The team defined its strategic objectives, which provided guidance for the team's activities. Team deliverables were defined to direct the team's work towards products or processes to meet the strategic objectives. Critical success indicators were defined to measure the effectiveness of the team. For example, one strategic objective was to reduce non-neutral wrist movements during product handling. A deliverable defined for this strategic objective was "To develop a human-machine interface that allows neutral wrist posture when loading 200mm wafer cassettes onto equipment." The critical success indicator was to have four interfaces designed within six months. Another strategic objective was "To define a robust decision-making process using documented engineering criteria." The deliverable was to have a decision-making process identified and engineering criteria defined. The critical success indicator was to have the decision-making process ratified by all sites in the first quarter of the team's charter. These are 2 examples of strategic objectives used at Intel during the VF ergonomics team's formation. Other objectives were developed to guide the team successfully through its first years.

4.2.2 Communication Process

As with any team process, communication is key to the success of the VF ergonomics team. The team addressed this important factor by defining a communication process that would ensure that all stakeholders were in alignment with the VF ergonomics team direction and decisions. This included the development of a team project roadmap and the development of a quarterly update process to provide management with the updates necessary for accurate guidance and direction of the team's work.

The team project roadmap was created to define and track progress of team projects. It was published and regularly updated on a web page for any stakeholder or customer to view. The project roadmap included details of the project, such as primary owner, ergonomics contact, project schedule and current status. The communication of project status and issues was very critical, since one VF ergonomics team member was providing ergonomics expertise for all sites on any given project. The VF ergonomics team reviewed the project roadmap monthly to ensure that all team members were aware of project status and could give input.

A quarterly update to management was defined to be a key communication strategy, since factory management was viewed as a very important stakeholder of the VF ergonomics team. The management team played a functional role in providing direction for the VF team's strategic objectives and ensuring they were in-line with Intel's corporate directives. During the quarterly updates to management, the VF ergonomics team reviewed progress against critical success indicators and the latest revision of the project roadmap.

4.2.3 Decision-Making Process

Establishing a sound decision-making process was imperative to create an effective team environment. For success in managing multiple projects across multiple sites, sound and objective decision-making criteria would need to be determined and applied consistently by the team. Therefore, the team set clear ergonomic guidelines based on engineering criteria for all solutions to meet or exceed. All ergonomics evaluations and risk assessments performed in the VF utilized these ergonomics engineering criteria, thereby facilitating sound and repeatable decision-making in the team environment.

The team adopted the "Disagree and Commit" method for reaching a final agreement with respect to issue or solution proposals. The concept of "Disagree and Commit" was employed when consensus could not be reached, and required that dissenters agree to support the majority opinion. There was also a time allotted for each team member to research the given issue and bring their final recommendation to the team for review and discussion. This allocation of the time for thorough investigation resulted in thoughtful and sound team decisions.

4.2.4 Solution Development

If the Virtual Factory ergonomics team determined that an ergonomic risk existed and required resolution, a team was formed to develop that solution. The solutions team consisted of all or some of the following members: VF ergonomics team representative, original equipment manufacturer (OEM) representative, and/or Intel engineering and manufacturing representatives who owned the tool or system needing modification. Like the VF ergonomics team member, the Intel engineer was part of a VF engineering team and represented all the Virtual Factory sites affected by the project.

Given this structure, the solutions team could include a VF ergonomics team representative from New Mexico, an Intel engineer from California, and an OEM representative from Montana. The solution developed by this solutions team could potentially be implemented in New Mexico, California, Oregon, Arizona, Ireland, and Israel. Therefore, it was critical that the VF ergonomics team member was regularly communicating project status and key decisions to the VF ergonomics team, and the Intel engineer was regularly communicating project status and key decisions to the VF engineering team. Excellent communication was key to the success of the project, since all sites would be required to implement the final solution whether or not they were part of the solutions team.

The solutions team was chartered to review the ergonomic risk, define multiple solutions to mitigate the risk, and ultimately determine the proposed solution. The solutions team then conducted adequate testing of the proposed solution, and ensured the final solution was implemented at all affected sites. The following sections describe the solution testing procedure used.

It is important to note that the primary leader of the solutions team was not a VF ergonomics team member, but was the VF engineering team member who owned the system being modified. The primary role of the VF ergonomics team member was to ensure that the solution met the established ergonomic guidelines.

4.2.5 Solution Testing and Implementation

The final proposed solution is taken to the Alpha test stage, which consists of prototype development and initial testing of the solution at one Intel site. Examples include: an engineering solution prototype developed at one site, or an administrative solution field tested at one site. Data are gathered at the Alpha test stage against engineering criteria and user surveys. Improvements are made to the design based on these data, and the project moves into the Beta test stage.

During the Beta test stage, the Alpha-stage solution is tested against the same engineering criteria and user surveys at a different Intel site. The use of two different sites during the Alpha and Beta test stage is important in order to minimize the possibility of a factory bias. From this Beta test, it is determined whether this solution passes or fails the criteria. If the solution fails the Beta test criteria, improvements are made and the solution undergoes a second Beta test. When the solution passes the Beta criteria, it is determined to be the solution to be implemented by the Virtual Factory at multiple sites.

The solutions team ensures that the affected VF sites implement the final solution. This requires each site to commit to a completion date, and the solutions team tracks performance against the committed schedule. This information is also tracked on the VF ergonomics team's project roadmap.

Any engineering solution is documented and communicated to the OEM with the expectation that all future generations and Intel purchases of this tool from the OEM will include this solution. This demonstrates the power of the VF ergonomics team approach: the process ensures that the risk will be resolved at multiple sites, and that any future factory using this equipment will use the same solution.

4.3 CHALLENGES

4.3.1 "Not Invented Here"

During introduction of a Virtual Factory solution process, Intel encountered resistance to the solutions created at one site and implemented in other sites. Intel commonly referred to this as the "not invented here" phenomenon. However, as people became accustomed to the VF ergonomics team process and to using the solutions developed at other sites, resistance to solutions became minimal. As trust in the Virtual Factory system increases, the process matures, and solutions are more easily reached and approved at the Virtual Factory level.

4.3.2 "What Defines an Ergonomic Risk?"

Agreeing on the criteria for risk, and determining when an issue needs a solution can be challenging in a VF ergonomics team approach. Input from multiple sites, different site focuses, and different opinions of the VF ergonomics team members can make it difficult to reach consensus. The critical element needed to alleviate this issue is an agreed-upon, documented set of ergonomics criteria, or Ergonomics Guidelines as they are called at Intel, that each site uses to determine ergonomic risk. This minimizes disagreement between ergonomics representatives and allows the VF ergonomics team to have a unified voice and a single document on which to base its decisions.

4.4 THE BENEFIT OF A VF ERGONOMICS TEAM APPROACH

Using the approach outlined above, Intel has made significant ergonomics improvements to factories throughout the world. Because of the team approach, Intel has minimized duplication of effort and cost, and maximized our safety systems. Using the VF ergonomics team approach for the last eight years, Intel has completed hundreds of ergonomics improvement projects affecting thousands of tools and improving the workplaces of tens of thousands of employees worldwide.

4.5 SOME KEY SUCCESSES

There are numerous examples of successful use of the VF ergonomics team approach at Intel. This paper will outline two examples for illustrative purposes. These examples by no means cover the entire scope of solutions implemented at Intel, but they illustrate the potential of the VF ergonomics team system.

4.5.1.1 An Engineering Solution using the VF Ergonomics Team Approach

A VF ergonomics team approach was used to implement an engineering solution, which eliminated non-neutral postures when loading product onto Intel factory equipment. Based on postural and force analysis, it was determined that the standard interface for loading product onto factory equipment required non-neutral hand and wrist postures and significant lateral forces (see Figure 4.1). The VF ergonomics team recognized that this combination of posture and force was a violation of our documented ergonomics criteria, and therefore a solution was needed.

Figure 4.1 Non-neutral postures and lateral force required to load product.

The VF ergonomics team understood that many of the production tools had the identified non-neutral interface problem, and Intel used a parallel approach to resolving the issue. For illustrative purposes, a description of one design effort will be outlined. A team was formed consisting of a VF ergonomics team member, several engineers related to the equipment being changed, and OEM representatives. The team defined several options for resolving the non-neutral interface and documented the possible solutions. A feasibility study was done on each option to determine engineering design requirements, cost, and schedule. A final solution consisting of an engineering modification to the tool load interface was selected (see Figure 4.2).

Figure 4.2 Engineering solution ensuring neutral hand and wrist postures.

The OEM was tasked with designing the engineering modification and presenting a prototype to Intel. Intel had defined ergonomics and engineering criteria for the OEM to meet, and this guided the prototype design effort. Once the prototype was designed, the OEM ran an Alpha test at their facility. This entailed testing the prototype design in various ways to ensure that it met the ergonomics and engineering criteria and could stand up to manufacturing conditions at Intel. Based on the data gathered during the Alpha test phase, the solution was modified and re-tested. Once the Alpha test prototype passed all the requirements, the design entered the Beta test phase.

During Beta test, one Intel site agreed to install the solution on a production tool for one month. Throughout the month data were gathered to ensure that all engineering and ergonomics criteria were met. In addition, survey data were collected from the factory to ensure that the solution met the needs of the users. If during the Beta test phase it was determined that any modifications to the design were required, the changes were made. The new design was re-tested by undergoing another Beta test phase. When the Beta test phase finished with no violations to the engineering and ergonomics criteria and no further modifications required, it was considered a successful Beta test. At that time the solution went into the proliferation phase.

During proliferation, the VF sites impacted by the solution agreed to implement the solution on all tools in their factories. A schedule was set for each factory to meet, and a monthly review of the schedule occurred to ensure timely implementation of the solution across the sites.

Many of the tools at Intel factories required this engineering solution to resolve a non-neutral load interface, and thousands of tools worldwide were upgraded with a similar engineering solution. In addition, Intel drove an industry standard to ensure that all future semiconductor tools were designed with a neutral load interface. This example demonstrates how a VF ergonomics team approach solves an engineering issue for Intel's current factories and ensures future tool purchases are delivered with the same solution.

4.5.1.2 An Administrative Solution using the VF Ergonomics Team Approach

A VF ergonomics team approach was used to implement an administrative control related to product handling methods at several Intel sites. Based on postural analysis, it was determined that some common methods used to handle product at Intel factories resulted in non-neutral postures of the hands and wrists (see Figure 4.3). The VF ergonomics team determined that the non-neutral postures were in violation of our documented engineering criteria, therefore a solution was needed. Investigations indicated that an engineering control to eliminate the non-neutral postures was not feasible, so an administrative control was pursued.

A team was formed consisting of a VF ergonomics team member and manufacturing team members who regularly handled product at an Intel factory. The team developed several alternate product-handling methods that would ensure the use of neutral hand and wrist postures. The team documented these methods in a training package.

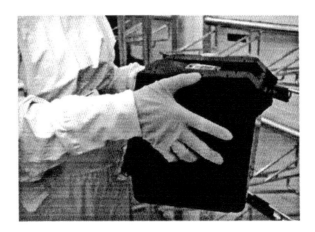

Figure 4.3 Non-neutral hand and wrist postures used to handle product.

An Alpha test factory was selected to test the alternate methods for one month. The team trained a sample of the factory population on the proposed methods, and the trainees agreed to try the alternate methods during regular factory operations. During that time, postural improvements were documented and survey feedback was collected on the methods. Based on the data collected during the Alpha test, changes were made to the methods. The modified methods were documented in a training package for use during Beta test at a different factory

A different Beta test factory was selected for further test of the improved methods. The same training, testing, and data collection methods were used during the Beta test as were used during the Alpha test. The data collected at the Beta test passed the VF ergonomics team's guidelines, and the product handling methods were ratified as a Virtual Factory solution. At this time, each factory committed to train and audit the use of the methods in their factory (see Figure 4.4).

Figure 4.4 Neutral postures for handling product with approved method.

As a result of this process, four product handling methods allowing neutral hand and wrist postures were approved and implemented at six factories. The number of employees impacted by this control is approximately 4000 worldwide. Audits have shown that the use of the improved methods is increasing at Intel factories, and user concerns and injuries related to product handling have been significantly reduced, since the implementation of the methods training.

4.6 CONCLUSIONS

Intel Corporation has utilized a Virtual Factory approach, making efficient use of ergonomics resources to resolve significant risk, and ensuring consistent ergonomics practices across all Intel sites. Intel's ergonomics injury trends have been decreasing steadily since the VF ergonomics team approach was created. Intel has found that a successful team approach to resolving ergonomics issues is one that includes a member from each factory site, joined together with common focus, objectives, criteria, and solutions. Through this approach, Intel's ergonomics program has achieved world class status.

CHAPTER FIVE

Deferential Analysis and Integrated Solutions
Ergonomic Re-engineering of a Sub-assembly Line

Boris Povlotsky, Dean Georgoff, and Gwen Malone

5.1 ABSTRACT

The word "deference" means to yield to the opinions of others. The concept behind the use of the word deference consititutes the premise for a new approach for generating the type of information exchange to achieve the goal.

This case study is intended to highlight an experience in deferential analysis and integrated solutions during the re-engineering of a sub-assembly line. Operators made many valuable suggestions during the re-engineering process, and facilitated a smooth transition to new procedures, tools, etc. Face to face surveys, aided by using a specially developed questionnaire, allowed engineers to optimize the assembly process, eliminate ergonomic risk factors, and improve both the efficiency and the quality of work.

This case study reflects "reactive ergonomics." However, the deferential analysis approach also lead to proactive ergonomic implementations. Ergonomic implementations can comply with the principles of lean production, including teamwork, communication, elimination of waste, and efficient use of human resources, facilities and materials. A significant result was that operators accepted almost 98% of the re-engineering.

This paper illustrates a challenge within the manufacturing environment, as well as what failed and what succeeded for ergonomic re-engineering.

5.2 INTRODUCTION

This project took place in a relatively new lean manufacturing facility that produces powertrain components. As it is with new plants, obtaining rated production output tends to require a maturity of its processes. This project needed to resolve a bottleneck at the beginning of the main assembly line. Actual output of the Case Sub-assembly operation was approximately 20 percent less than expected. Several investigations turned up a wide array of opinions and proposed solutions to the bottleneck, each of which seemed to be incongruent with the others. There was no apparent single issue that alone accounted for the difference. Each proposal focused on one aspect of the problem. Often the source as well as the corrective action proposed by the various engineers, managers, and specialists seemed to be dependent on their respective disciplines. For example, department supervisors claimed that "social loafing" was the problem; engineers recommended costly machinery and equipment changes; operators suggested that there was not enough manpower assigned; and the union's ergonomics representative cited fatigue related to a poorly designed process as the fault.

Management was concerned with the need to invest in capital equipment if the operators had time to perform their individual tasks. This perspective favored an administrative solution over an engineered one. A time study revealed that the three operators involved, though independently capable of performing their tasks at the rated output, could not maintain the synchronization required, delaying the flow of parts through the department. These minute work stoppages multiplied and accrued throughout the shift.

What early on appeared to be "social loafing" was a natural attribute stemming from a worker's idle moments waiting for another operator's work output to enable them to continue working. Understanding this, attention shifted away from administrative solutions and toward undocumented work that had escaped the established standard time analysis. It was our belief that industrial engineering, process engineering, and ergonomics engineering needed to be integrated. A detailed evaluation was needed to allow small problems requiring small solutions to surface.

This simplistic method of evaluation and problem-solving later confirmed that the quality of ergonomics was an important factor in the department's throughput.

5.3 THE DEFERENTIAL ANALYSIS AND INTEGRATED SOLUTIONS APPROACH

The deferential analysis and integrated solutions approach in this case study can be regarded as a result of adhering to concepts that we, the co-authors regarded as important theories of innovation within a socio-technical system (Duffy, 1999) or environment. Progress toward producing productive solutions within the socio-technical systems of both the plant and the department was blocked by a basic lack of synergy among the people involved.

Another concept that contributed to the construction of the deferential analysis and integrated solutions approach was to equate Lean Production (Wormack, 1990), Industrial Engineering (Shingo, 1989), and Ergonomics (Hancock, 1999). The work of this project was based on the premise that minimizing any effort required to perform a job benefits both the ergonomics and the productivity.

A survey was implemented to gather information from the operators who actually performed the work. Their perception of the tasks that were most difficult and caused discomfort, slow-down, and quality concerns were considered in making final decisions. Also, a more constructive exchange between the engineers and the operators was initiated. Our willingness to alter decisions based on a continuous stream of new insights introduced a number of iterations to the re-engineering process and produced a widely accepted end result.

5.4 STATUS QUO AND CHALLENGE

The main assembly line is balanced to produce 78 units per hour, given appropriate operation of all the sub-assembly lines that supply it. However, the Case Sub-assembly line was averaging only 65 units per hour, despite standard element times for each operation that were set to meet the 78 unit-per-hour demand. Therefore, the sub-assembly line operation had to run overtime and weekends to compensate for throughput problems that were not readily identified.

5.4.1 Operators

There is a team of four operators per shift and two 8-hour shifts per day for the Case Sub-assembly line operation. Operators are rotated every 90 minutes to each of the other operator's tasks within the Case Sub-assembly line. Members of both shifts could be described as a diverse group of union workers, demographically different by age and seniority. Seniority ranged from two months to more than 25 years. Ages ranged from 21 to 53. Stature ranged from 5th to 95th percentile of the North American female and male populations. Both shifts included 20% women. Some workers were left-handed. Some of operators were working in an assembly job for the first time. Each individual developed his or her own personal pace for each operation, providing performance speeds that varied from slow to quick. In some circumstances, due to operator rotations, "quick" performing operators were forced to work at the pace of the operators that were "slower."

To compensate for the range of all the operator's anthropometry and performance speed, it seemed ideal to implement adjustable workstations for every individual (Povlotsky, 1986) of this diverse group. But this would have meant purchasing all new equipment designed with complex adjustment features.

In considering the Case Sub-assembly line as a socio-technical system, it became obvious that we should anticipate some obstacles and difficulties such as:
- Natural individual resistance to change.
- Different goals of individuals, union and administration.
- Quick fix syndrome.
- Budget.
- Deadlines for project completion.

5.4.2 Status of Old Operation

Six generalized tasks comprise the work of the four operators in the department. The layout for the old operation is shown in Figure 5.1.

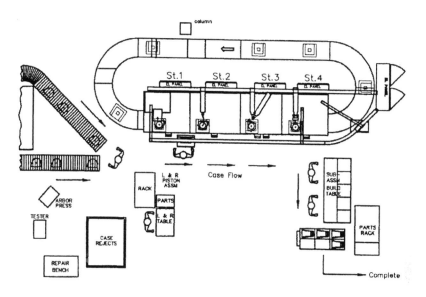

Figure 5.1 Old Layout Operation.

Task 1.

The operation No. 1 (stations #1-#4) started with an operator 1 manually picking up a case from one of two conveyors supplied by an upstream washing process, turning it 180 degrees and then installing it into a pallet. It is important to note that successful part flow required that the operator alternate between the supply conveyors. This operator then proceeded to prepare each of four cases and four automated machine stations (Figure 5.1, St.1-St.4) linked by a powered conveyor for cycle by loading small parts to various fixtures and rotating pallets into proper position for cycling. As the operator returned to the start of the process, the cycling of the four machines had been initiated by activating individual start switches.

Task 2.

Operator 2 rotated the pallet and manually transferred the case to a flat- top table and upon completion of his task, the case would be pushed to operator 3. This point of transfer was a source of bottleneck because of the close interface between the operators 2 and 3. Inadequate space between them resulted in the lack of a buffer, and consequently a leader-pace situation. After the transfer, the operator 3 would continue installing more components and then manually placing the finished case onto a cart for transport to the main assembly line.

Tasks 3 through 6.

Operator 4 performed tasks No. 3 through No. 6, which supported the other three operators by pre-assembling several small parts, stocking parts bins, and repairing parts rejected from a leak test process. This operation was self-paced.

5.4.3 Objective and a New Approach

Deferential analysis and integrated solution strategies were initiated to encompass all possible socio-technical aspects of the Case Sub-assembly line. Therefore, the objective for this strategy was to integrate aspects of ergonomics, lean production and industrial engineering to:

- Eliminate and/or reduce ergonomic risk factors.
- Reduce case lifting by operators.
- Implement adjustable workstations.
- Improve work flow and layout.
- Improve tools and procedures.
- Make efficient use of team resources.
- Eliminate waste.
- Balance the Case Sub-assembly line tasks.
- Promote teamwork and communication.

5.5 IMPLEMENTATIONS

Deferential analysis and integrated solutions constitute an iterative process. A number of proposals were evaluated and altered with new versions until the re-engineering concept met consensus of the implementers. This section highlights what occurred as a result of the exchanges between the engineers and operators. A "face to face" survey in the form of a specially developed "Workplace Analysis" questionnaire was used for the integration of several sources of old and new data regarding the tasks. The old data, such as symptoms questionnaires, risk factor checklists, medical records, time motion study, and work flow analysis lacked the clarity that the operators could provide regarding problematic tasks, tools or methods.

5.5.1 "Workplace Analysis" Questionnaire

One of the tools successfully used during deferential analysis of operator workstations and working procedures was the specially developed "Workplace Analysis" questionnaire.

Adding the information gathered by the "face-to-face" survey process and documented on the "Workplace Analysis" questionnaire enabled the engineers to gain a perspective for problem-solving that was directly linked to all aspects of a specific worker's tasks. It was more difficult to make conclusions when operators were rotated, executing a number of different operations.

Every question in our specially developed questionnaire relates a specific operation, tool, or physical complaint associated with a working procedure. Furthermore, the "face-to-face" survey was an opportunity to "educate" respondents in ergonomics, establish a working team, and "educate" the investigator in the details of the operator's work.

The following are samples of the "Workplace Analysis" Questionnaire used in our survey. Figure 5.2 is a sample of questions related to operators' data and difficult tasks they performed.

WORKPLACE/WORK PROCEDURE ANALYSIS

CASE SUB-LINE

> THE GOAL OF WORKPLACE ANALYSIS IS TO ESTABLISH A JOINT TEAM CREATIVE ENVIRONMENT THAT WILL MAKE YOUR WORK MORE SAFE AND PRODUCTIVE .

Date__/__/__

1. OPERATOR
  Name:_____Height:__feet__inches.

  Dominant hand: Left [], Right []

2. JOB INFORMATION

2.1 Job Occupation_____ Shift: First [] Second []
          How long have you performed this Job?___Weeks___Months___Years

2.2 "Popular" Job Title: *Sub-line* _____ *Build table*_____*Piston Assm* ___

3. WHAT IS THE MOST DIFFICULT PROCEDURES AND WHAT KIND OF IMPROVEMENTS (such as tooling, facility, layout, operation, etc.) WOULD YOU SUGGEST FOR THE FOLLOWING AREAS?

3.1   Cases handling from dryers' conveyors:
The most difficult procedures _____
Suggested Improvements _____
*Comments*_____

3.2 Machine Sub-Assembly four stations:
The most difficult procedures _____
Suggested Improvements _____
*Comments*_____

3.3 Case Sub-Assembly build table:
The most difficult procedures _____
Suggested Improvements _____
*Comments*_____

3.4 Rod & Lever Assembly table:
The most difficult procedures _____
Suggested Improvements _____
*Comments*_____

3.5 Leakers/Repair :
The most difficult procedures _____
Suggested Improvements _____
*Comments*_____

Figure 5.2 Sample of questions related to operator data and difficult tasks.

Figure 5.3 illustrates how operators were asked to indicate which task they believed caused a particular discomfort and/or pain that they had been experiencing.

Some of the questions may be familiar because they reflect key points selected from other well-known samples of checklists or symptoms questionnaires.

4. IS THERE A SPECIFIC WORK TASK OR ACTIVITY AT THE FOLLOWING AREAS, THAT YOU CONSIDER TO BE THE CAUSE OF YOUR PAIN OR DISCOMFORT (INCLUDING WEAKNESS, STIFFNESS, TINGLING, ETC.) THAT YOU FIRST NOTICED DURING THE PAST 12 MONTHS THAT LASTED LONGER THAN 7 DAYS?
If yes, please shade the area, front and /or back, on the drawing below, where discomfort or pain occurs
Write down the number that indicates the severity of the symptoms based on the following scale.
1-Mild   2-Moderate   3- Severe

4.1 Cases handling from dryers' conveyors:

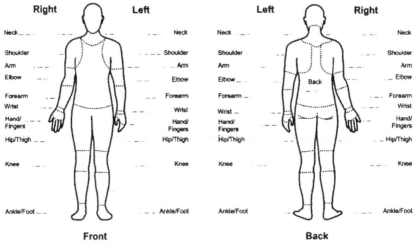

4.2 Machine Sub-Assembly four stations:

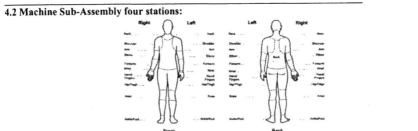

4.3 Case Sub-Assembly at build table:
4.4 Rod & Lever Assembly table:
4.5 Leakers/Repair :
4.6 L&P Piston Assembly Table:

Figure 5.3 Sample of questions related to operator discomforts and/or pain.

5.5.2 Radical Solution

The proposal made just prior to the deferential analysis demonstrated the "quick fix" mentality; it focused on resolving only the problem of operators 2 and 3 working at a single bench by creating a buffer space between them. A decision was made to halt implementation of this proposal. To continue would have reflected the lack of complete information. Several other factors, later understood to be critical to making improvements were not addressed in this approach. Deferential analysis greatly improved the development of very productive re-engineering plans.

The analysis led to the fundamental decision that two build tables were essential, and the final locations should be on the opposite side of the Case Sub-assembly machine conveyor.

The operators were adamant that part flow through the operation had to be left to right. Aside from satisfying the end users, we discovered several other benefits of this flow direction that otherwise would have been missed. For example, the parts supplied to the operators could be stocked from the same direction, and the flow of case racks in and out of the area was improved in terms of the plants layout. This more substantial and complete solution was achieved from analysis of several different options shown in Figure 5.4 and discussed below (Options "a" to "d").

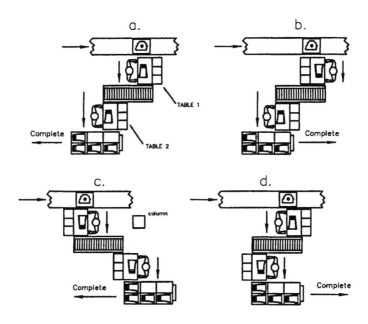

Figure 5.4 Options for a Radical Solution.

Option "a" – This option, which was finally selected for implementation, offered the stereotypical left-to-right part flow that the operators had preferred. Component parts are provided to each operator via flow-through racks in front of their workspace (Figure 5.5, build tables 1 & 2). Each of the racks is accessible to material handling personnel without interference with the conveyor. This arrangement allows for the elimination of a storage shelf in the department, and some tasks previously done by the operator can be assigned to stocking duties.

Option "b" – Left-to-right part flow is not maintained for the first operator. Also, the flow-through racks are hard to access with the loading points on opposite sides of the conveyor.

Option "c" – Aside from violating the left to right part flow for both operators a column interferes with proper access to the first operator's work area.

Option "d" – This option has the same faults as option "b" except that the second operator is forced to work from right to left, instead of the first operator.

5.5.3 Re-arrangement of Operators Assigned to the Team

Two operators, who were assigned to the upstream operations not previously described, became part of this more comprehensive solution. Each ran identical leak test machines, and loaded the wash machines that supplied the cases to the subassembly line. These two operators and their respective tasks were combined with the four sub-assembly line operators. This change enabled some work to be transferred from one of the bench operators to two leak test operators, achieving better line balance.

5.5.4 One Conveyor Versus Two

Enlarging the team from 4 operators to 6 operators revealed more about the consistent flow of material and how it related to ergonomics. The importance of the existing two conveyor lines as a critical link and major contributor to ergonomic risk factors became apparent. Figure 5.1 shows two conveyors feeding the first operation of the subassembly line. The operator removing the cases from the two supply conveyors would naturally select the nearest conveyor more often. As a result, the upstream operation supplying the far side conveyor would be blocked, forcing the worker either to stop or to carry cases to the other washer. Also, the cases moved to the end of the original conveyor upside down and at least a step away from the Case Sub-assembly line load point. The new conveyor, with two 90 degree turning mechanisms, merged the two supply lines into one unload point, and turned the case over to the orientation used in the next step of the process (Figures 5.5 and 5.6). With the new conveyor, the flow of cases had improved by providing a single delivery point. Also, the case's orientation and location eliminated the need for the operator to turn the case over manually, and minimized the distance to its destination.

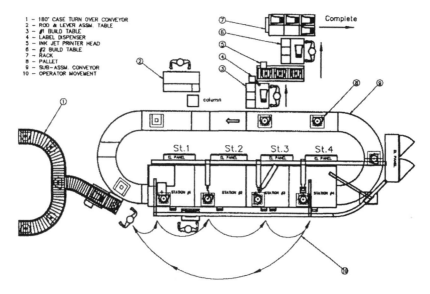

1 – 180° CASE TURN OVER CONVEYOR
2 – ROD & LEVER ASSM. TABLE
3 – #1 BUILD TABLE
4 – LABEL DISPENSER
5 – INK JET PRINTER HEAD
6 – #2 BUILD TABLE
7 – RACK
8 – PALLET
9 – SUB-ASSM. CONVEYOR
10 – OPERATOR MOVEMENT

Figure 5.5 New Layout.

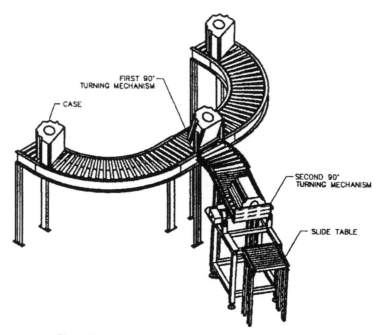

Figure 5.6 Conveyor with two 90° Case Turnover Mechanisms.

5.5.5 Project Momentum

During this project several other important changes were made or planned, including a specially-designed electrically-adjustable assembly table, pallet-turning mechanisms at several stations, a pop-up device to remove the case from the pallet easily, a label peeling mechanism, and an inkjet printer for part numbering.

Although the initial project has been completed, the ongoing process of re-engineering continues to create further improvements.

5.6 CONCLUSION

Based on efficiency improvements alone, this project could be deemed successful. The department is now capable of producing the rated volume daily, representing a 15 percent improvement. This accomplishment was largely due to eliminating or reducing the difficulty found in several small tasks. This series of small, integrated solutions equated to ergonomic benefits without costly interventions.

A key factor in discovering a wealth of improvements within this ergonomics re-engineering opportunity was the deferential analysis approach that was used. It was not until we changed our perspective from one of a series of independent machines/operations to that of a socio-technical system, complete with an array of human and technical interactions, that we began to make progress.

We would suggest that the concepts presented in this case study could be applied in a proactive fashion to ensure good ergonomic conditions. One measure of success is the level of acceptance of change. Often ergonomic solutions fail because they are not accepted by those that they are intended to help. The reason that such potential solutions fail is that they are derived in isolation.

Integration of all sources of knowledge and information, especially worker's experiences, during the problem-solving process leads to well-integrated solutions.

5.7. REFERENCES

Duffy, V.G. and Salvendy, G., 1999. The impact of organizational ergonomics on work effectiveness: with special reference to concurrent engineering in manufacturing industries. *Ergonomics*, 92(4):614-637.
Hancock, P.A., (Ed.), 1999. *Human Performance and Ergonomics*, Academic Press.
Povlotsky, B., 1986. Workstation Customized for Productivity, Health and Safety. In W. Karwowski, (Ed.) Trends in Ergonomics Human Factors III, (Amsterdam, The Netherlands: North-Holland), pp. 309-313.
Shingo Shigeo, 1989. A Study of the Toyota Production System From an Industrial Engineering Viewpoint (rev.), Cambridge: Productivity Press.
Wormack, J.P., Jones, D.T., Roos D., 1990. The Machine That Changed the World. New York: Harper Perennial.

CHAPTER SIX

Whirlpool Corporation Clyde Division Ergonomic Process Management Written Plan

Wendell H. Howell, CSP, ARM, WSO-CSM,
& Ergonomic Process Management Team

6.1 INTRODUCTION

The Clyde Division of the Whirlpool Corporation is the world's largest manufacturer of automatic clothes washing machines, employing 3,800 persons in a 2 million square foot facility located in Northern Ohio.

Our ergonomic challenges come as a result of our daily production of 19,000 units in a sometimes hand intensive production environment. Safety and quality are top priorities at Whirlpool, demanding that we address ergonomic issues as an everyday requirement.

6.2 ERGONOMIC PROCESS MANAGEMENT PLAN

We have approached these ergonomic challenges by creating a formal comprehensive process designed to identify ergonomic problems, determine practical solutions, and implement them in a consistent and effective manner.

A written plan forms the foundation for the Clyde Division's ergonomics activities. This written plan is included as the Appendix.

There are six basic elements that form the foundation of our ergonomics program: (1) Leadership Involvement and Employee Participation; (2) Hazard Identification; (3) Job Hazard Analysis and Control; (4) Training; (5) Medical Advisory and Case Management; and (6) Process Evaluation.

Leadership involvement is the key to the Clyde Division ergonomic program success. None of the other elements would be effective without the active and visible support of our Senior Management. They have provided the resources for ergonomic improvements and are actively involved, many of them having participated in the same training as other ergonomic team members.

Hazard identification and job hazard analysis are accomplished through the Peer Audit Safety Teams (PAST) and the Ergonomic Process Management (EPM) team. Both groups are comprised of a cross section of affected hourly and salaried employees. The EPM team has a Mission Statement, goals and a targeted selection of team members, who address ergonomic concerns in a systematic manner. EPM member roles are clearly identified, as well as the process to fill vacancies.

Formal classroom training is provided for PAST and EPM members. All employees receive annual training.

Our EPM Committee includes an Occupational Health Care Physician and a Certified Occupational Health Nurse (COHN), providing medical advisory and case management support.

An informal ongoing evaluation of the EPM and PAST activities is conducted as well as an annual evaluation of the EPM.

The major ergonomic efforts are attributable to the PASTs. PASTs conduct most of the actual ergonomic worksite analyses. If they are unable to develop a solution, the Ergonomic Process Management team will address the issue. A defined approval process is used and coupled with a planned implementation strategy. The solution will be evaluated in real time with feedback from the users who have in most cases had input into the solution. Training is provided for PAST members through 2-day ergonomic seminars conducted on-site by the Joyce Institute (now A.D. Little). The knowledge gained by the PASTs has assisted in identifying and resolving numerous ergonomic issues. Manufacturing departments, called Component Manufacturing Cells (CMCs), have yearly goals on the number of ergonomic worksite analyses their PASTs will accomplish.

For consistent production of a quality product safely at our very high production levels, we have employed many engineering interventions. These include articulating arms for hand tools that remove or reduce all vibration and torque. Tilting tote stands are common; and some can be rotated and/or raised as necessary. Manual material handling is avoided whenever possible. Where it has been feasible to do so, manipulators that pick up cumbersome and heavy parts are in use on current production lines and are being installed on our new lines. In-house designed ergonomic assists are now common. Conveyors are at prescribed heights.

Even with these interventions, we also employ rotation of employees in cells with 3 to 23 employees. Where engineering controls are not currently practical or totally effective, planned rotation provides relief to assembly personnel. Rotation times vary from 30 minutes to an hour.

The ergonomic process includes Project Engineers working with the Safety department at the concept stage of revising or purchasing new machinery, equipment and processes. Reviews are included at key points in the process, until the equipment is installed and in full production.

6.3 SUMMARY

The Clyde Division of Whirlpool feels it has a successful ergonomics program, which involves employees at all levels and utilizes a multitude of intervention strategies. The framework for this program is supplied by a written plan which formalizes the program and provides a map to direct the program's activities. This written plan is included in this paper as the Appendix, to present an example of what has been found to be effective in this production environment.

Appendix

Whirlpool Corporation Clyde Division Ergonomic Process Management Written Plan

STEP I: MISSION STATEMENT

We will improve the quality of life for all employees, reduce related injuries and illnesses as well as associated costs and enhance productivity, reduce scrap/re-work and improve quality through the use of ergonomic principles in the workplace.

STEP 2: GOALS

Our goals are to:
1). Employ a systematic proactive, as well as reactive, process to reduce ergonomically related injuries and illnesses.
2). Reduce the negative impact on productivity, quality and scrap/rework.
3). Reduce employee pain and suffering due to ergonomic injuries.
4). Reduce related workers' compensation costs.
5). Improve employee morale by including them early in the process and by showing them the benefits of their efforts in the process.

STEP 3: ERGONOMICS PROCESS MANAGEMENT TEAM PERSONNEL

The Ergonomics Process Management team will provide the technical expertise required for the implementation of the ergonomics process management effort. The current Ergonomics Process Management team is composed of the following:
- Peer Audit Safety Teams (PAST)
- Product Engineering
- Manufacturing Engineering
- Benefits/Compensation Personnel
- Advisors
- Contract Employee Representative
- Maintenance/Skilled Trades

- Plant Safety Personnel
- Health Care Professionals
- Rehabilitation Employer Representative
- State of Ohio Bureau of Workers' Compensation - Division of Safety and Hygiene

STEP 4: TEAM MEMBER ROLES AND RESPONSIBILITIES

The Ergonomics Process Management team will typically meet on the third Thursday of each month at 1:00 PM in the Plant 3 Conference Room 3A. This may vary depending on circumstances. There may be meetings held more often than once a month if determined necessary by the team. The meetings will have specific agenda items to cover and action items will be identified and documented via minutes. If at any time any of these designated people wish to relinquish their duties, they will bring it up to the team and new people will be designated.

- Leader - There will be a Team Leader to facilitate the meeting.
- Recorder - The meetings will be documented via minutes generated by the recorder.
- Scribe - There will also be a scribe to record information on an overhead or flip chart, when necessary.

All information collected by the Team will be kept in the Team Leader's office.
Any Team member that has information pertinent to the Team should give it to the Team Leader.

The Written Ergonomics Plan, the meeting minutes and any other relevant information will be posted on the Intranet by a designated team member for any employee within Whirlpool to review. The meeting minutes will be sent via CCMail to all team members.

Team members should inform the Team Leader, prior to any meeting if unable to attend the meeting. If frequent absences occur, the Team will discuss future membership status.

STEP 5: FILLING VACANT ERGONOMIC TEAM POSITIONS

As a vacancy occurs within the Team, the Team Leader will contact the Manager over the area from which the person was from to get a replacement for the vacating person.

STEP 6: TRAINING AND EDUCATION

The purpose of training and education is to ensure that employees are sufficiently informed about the ergonomic hazards to which they may be exposed and thus are able to participate actively in their own protection. Training allows Managers, Advisors and employees to understand ergonomic hazards associated with a job or production process, their prevention and control and their medical consequences.

New Ergonomics Process Management Team Member Training:

New members to the Ergonomics Process Management team will receive in-depth training as soon as possible after joining the Team. An assessment will be made on a quarterly basis to determine if a training session should be scheduled.

The veteran Team members and/or Bureau of Workers' Compensation/Division of Safety and Hygiene will provide the training. The training will cover at a minimum the following topics:

- Definition of Ergonomics
- Applications of Ergonomics (e.g., Manual Materials Handling, Visual Display Terminals, etc.)
- Types of injuries and illnesses (e.g., Cumulative Trauma Disorders)
- Injury/Illness and workers' compensation statistics (e.g., cost per unit, recordable, etc.)
- Basic anatomy
- Risk factor identification
- Anthropometry
- Control measure formulation

The materials used to provide the training may include reference book(s), videos, overheads, slides and handouts.

Continual refresher training for all Ergonomic Team members will occur on an ongoing basis by having vendors present various products and other topics discussed during Team meetings.

General Awareness Ergonomics Training For All Employees:

The general awareness training for all employees will provide an overview of general ergonomic principles and also instructions on how to apply the principles to their work areas. The Ergonomics Process Management Written Plan will be reviewed and a handout of key points provided so all employees will know how to get involved with the process.

General awareness training will be conducted annually. The training will consist of a video and a key points handout.

New Hire Training:

All new hires will receive ergonomics training as part of the orientation training administered by the Peer Audit Safety Team(s) prior to starting work.

The new hire training will provide an overview of general ergonomic principles and also instructions on how to apply the principles to their work area. The Ergonomics Process Management Written Plan will be reviewed and a handout of key points provided to all employees will know how to get involved with the process.

Worksite Analysis:

The Engineers for the work area being modified will provide training for affected employees regarding the specific worksite modifications. This training will begin before any changes are made to a work area and will continue until all modifications are in place and tried by all affected employees.

Hands-on shadowing will be the preferred method of training with videos, handouts and vendor support used to supplement the training.

STEP 7: TREND DATA COLLECTION AND ANALYSIS

The Ergonomics Process Management team recognizes the importance of collecting and analyzing various types of data to be successful with their mission. In order for the Ergonomics Process Management team to be able to identify where injury/illness trends are occurring, the appropriate data must be collected and reviewed. This data can also provide a means to evaluate the implementation of the ergonomics process and to monitor progress accomplished.

Injury/Illness Records:

The Workers' Compensation representative on the Team will collect all injury/illness information and provide it to the Team on a monthly basis unless an injury demands immediate ergonomic analysis. The following are the types of records that will be provided to the Team:
- OSHA 200 and 101 (Ergonomics Issues Log)
- First Aid Log
- Claims/Workers' Compensation information

The Ergonomics Process Management team will conduct ergonomic trend analyses of the injury/illness data (e.g., reports of strains/sprains, over exertions and the Ergonomics Issues Log, etc.).

Production Records:

Advisors are responsible for gathering the Monthly Production Reports that are generated by the Information Technology (IT) department and for providing the information to the Team on a monthly basis.

The Ergonomics Process Management team will conduct an ergonomic trend analyses of the production related information (e.g., bottlenecks, low quality areas, high scrap areas, etc.).

The Floor Inspector will provide an ergonomics checklist to be completed when assessing a scrapped product. This checklist will be given to the Supervisor to be given to the Team as part of the monthly report.

Absentee Records:

The Health Center will conduct an ergonomic trend analysis of the absentee related information to determine what departments have consistent high absenteeism.

Employee Suggestions:

The Ergonomics Suggestion form will be used for employees to provide suggestions for improving a work area from an ergonomic perspective. This form can be obtained from all Supervisors who will in turn give the completed forms back to the Team Leader.

Once the form has been reviewed by the Ergonomics Process Management team, a copy of the form with the recommended action identified will be given to the employee who submitted the form and a copy will be retained by the Ergonomics Process Management team.

Ergonomics suggestions will be tracked via the meeting minutes and a quarterly review of all suggestions will be conducted to check on the status of the suggestions as well as review the Ergonomic Process Management team meeting minutes for updates.

Employees may also provide suggestions through the Peer Audit Safety Team(s) and/or by attending the Ergonomic Process Management team meetings.

STEP 8: PRIORITIZE JOBS

Once all data has been collected and analyzed, the Team will assign priorities to determine which jobs are evaluated first based on the following criteria:
- Frequency of injuries/illnesses
- Severity of injuries/illnesses
- Most people affected
- Easy to implement solution/modification
- Most potential for risk factor reduction
- Employee suggestion
- Peer Audit Team's Ergonomic Worksite Analyses with high priority

STEP 9: ERGONOMIC WORKSITE ANALYSIS

The Ergonomics Process Management team will provide information to the PAST teams on where ergonomic analyses may be needed. The PAST team will conduct the ergonomics worksite analysis using the Joyce Institute analysis tool. When the PAST team cannot perform an analysis or when assistance is requested, the Ergonomics Process Management team may conduct or assign the analysis.

A PAST team member(s) will videotape the jobs targeted for the ergonomics evaluation and complete the necessary analysis forms. These forms will be given to the Team Leader for tracking purposes. A PAST team member or the Team Leader will inform the Advisor of the affected area that the job will be evaluated. The Advisor of the affected area will inform employees that the job will be evaluated.

STEP 10: SOLUTION DEVELOPMENT

Once ergonomic risk factors are identified through the systematic worksite analysis discussed previously, the next step is to generate possible solutions/recommendations/interventions to reduce/eliminate the identified stressors.

The Ergonomics Process Management team and/or the PAST team and affected employees in the area (workers, Supervisors, Engineers), will brainstorm possible solutions. The PAST team member or the Ergonomics Process Management team member will document the brainstorming ideas on the worksite analysis form.

A cost/benefit analysis will be conducted on the solutions with the least cost yielding the most risk factor reduction receiving highest priority. The recommendations may include short term or temporary solutions with long term solutions identified as well. The solution will be a value-added solution.

STEP 11: APPROVAL PROCESS

The Ergonomics Process Management team and/or PAST team will provide the following to the Engineer responsible for the worksite modification: prior injury statistics, including workers compensation costs when known, and a recommendation for a follow-up procedure. The Engineer must present the proposal through the C2C process (when necessary). If the proposal is denied, the Engineer will meet with the team to determine which re-proposal is warranted.

STEP 12: SOLUTION IMPLEMENTATION

It is vital to the success of the ergonomics process that once final approval is granted for a proposed project, an implementation plan and evaluation strategy is developed and followed.

The Engineer will be responsible for seeing that this process is followed to project completion. The Engineer and Supervisor will meet with the affected employees to discuss the upcoming modification.

STEP 13: SOLUTION EVALUATION

Once a solution has been implemented, the Ergonomics Process Management team must ensure the solution did in fact accomplish what was intended (e.g., risk factor reduction). Therefore, every solution implemented must be evaluated to ensure employees satisfaction with the changes.

The Advisor of the affected area will interview employees within the first couple of weeks of the modification to receive feedback and report this information to the PAST team who will document the comments on the ergonomics worksite analysis form. After 6 months and after 12 months an ergonomic worksite analysis will be conducted again, ideally by the same person(s) who conducted the first analysis, and injury/illness, etc. statistics will be reviewed to determine if the modification was a success. All Ergonomics Process Management team members and PAST team members are encouraged to solicit feedback from workers regarding any ergonomics changes made to their work area.

STEP 14: MEDICAL MANAGEMENT

An effective medical management program is essential to the success of the ergonomics process. Health care providers must be interacted with routinely to exchange important information to help prevent and treat injuries. The elements of a medical management program include early intervention, evaluation and treatment of signs and symptoms and to aid in their prevention.

The Workers' Compensation Benefits representative is responsible for all aspects of the injury reporting, including first aid. The representative will provide a monthly report to the Team regarding that month's injury/illness activity. Through general ergonomics training, each employee will be encouraged to report early signs of symptoms/pain to the Health Center for appropriate action to be taken. The established Return to Work policy will be followed to get workers back to work as soon as possible.

STEP 15: PROCESS MONITORING

In implementing any process, procedures must be in place to monitor or track the progress of the process. The process must be reviewed on a periodic basis to evaluate the success in meeting the established goals and objectives. The results of the periodic evaluations will be shared with the PAST teams.

Annual summary of injury reports, production records, quality records, number of ergonomic worksite analyses conducted and solutions implemented, as well as employee suggestions received and followed through, will be reviewed to determine if an impact has been realized.

The Ergonomics Process Management team will review the Ergonomics Process Management Written Plan on an annual basis to determine if the goals have been accomplished. The Team will also ensure the methods established to implement the Ergonomics Process are still accurate; if not, the team will make appropriate changes and inform employees of the changes on the process.

DEFINITIONS

- Ergonomics - The study of the relationship between a worker and the workers' job. The objective of ergonomics is to adapt the job and workplace by designing tasks, work stations, tools and equipment that are within the workers' physical capabilities.

- Ergonomics Injury Management Process - The name of the process implemented to reduce the frequency of injuries and the cost of workers' claims by focusing on sound ergonomic principles and handling injuries in the most effective and efficient manner possible.

- Ergonomic Risk Factors - Conditions of a job, process or operation that may contribute to the risk of developing a cumulative trauma disorder. Examples of ergonomic risk factors include:
  * Repetitive motion
  * Forceful exertions
  * Sustained and/or awkward postures
  * Duration of activity
  * Vibration
  * Environmental factors such as extreme temperatures

- Cumulative Trauma Disorders (CTDs) - Disorders of the musculoskeletal and nervous systems which may be caused or aggravated by repetitive motions, forceful exertions, sustained or awkward postures, direct pressure, vibration or extreme temperatures. Some common CTDs are: tendonitis, tenosynovitis, DeQuervain's disease, trigger finger and carpal tunnel syndrome.

- OSHA 200 Log - A form developed by the Occupational Safety and Health Administration (OSHA) on which employers are required to list all work related injuries and illnesses except those requiring first aid only. Employers are required to post the previous year's OSHA 200 Log during the month of February for all employees to view.

- Interventions - Recommendations for reducing/eliminating the risk factors present at a particular work area. Recommendations and solutions are terms used interchangeably for interventions.

- Task Analysis - A method used to analyze a job to determine if ergonomic risk factors are present. The job is broken down into a number of tasks to be analyzed.

CHAPTER SEVEN

Using Human Factors To Understand and Improve Machine Safeguarding

Paul S. Adams

7.1 ABSTRACT

Maintaining the integrity of machine safeguarding systems has long been problematic for safety professionals and supervisors. Workers occasionally compromise safeguarding by defeating interlocks, cutting holes to gain access, or simply leaving guards off. These deliberate "unsafe acts" result from decisions by workers based on their individual "cost/benefit" evaluation. Human factors and behavioral psychology can be used to explain this decision phenomenon in terms of a decision criterion and man's limited ability to assess risk. Using this information, the ergonomist can increase the perceived "cost" and reduce the "benefit" of compromising the safeguarding system by applying a variety of design techniques. Among the techniques discussed are: making guards easy to replace, reducing visual obstruction, preventing inadvertent trips of emergency stops, and encouraging machine lockout.

7.2 INTRODUCTION

One enduring problem for safety practitioners is maintaining the integrity of machine safeguarding systems. Workers often compromise guarding by defeating interlocks, disabling emergency stop controls, cutting holes to gain access, failing to report and correct accidental damage, or simply leaving guards off. In addition, poorly designed guards can themselves present hazards (e.g., guards can create pinch points or pose a lifting hazard due to their size and weight). While human factors can play an important role in addressing both of these problems, this paper primarily addresses the role of ergonomic design in maintaining the integrity of safeguarding systems. Workers often cite production requirements, job pressure or stress, and ignorance as reasons for compromising or bypassing safeguards (Johnson, 1999), and one can argue about the root causes of such unsafe acts. However, applying human factors to safeguard design can effectively eliminate such behaviors.

7.3 OPERATOR DECISION-MAKING

Classical decision theory presumes that a rational decision-maker thinks logically about the decision. Various decision rule models have been proposed, although most adhere to what can be described generically as a cost/benefit approach. (See Lehto, 1997, for an overview of Classical Decision Theory.) Cost/benefit models suggest that humans choose deliberate behaviors after weighing the expected negative costs or consequences of the alternatives against the expected benefits of those same alternatives. Both expected costs and expected benefits are functions of the perceived probability of realizing an outcome, and the anticipated value of that outcome. Rational people choose deliberate behaviors based on internal criteria (i.e., their assessment of the overall net expected result and their willingness to accept risk).

Unfortunately, when it comes to assessing accident risk, humans are sub-optimal decision-makers (Wickens, 1984). They tend to be biased (Nisbett and Ross, 1980), have poor ability to assess probabilities (Zimolong, 1985), inaccurately assign cost or benefit values, for example the risk of accident (Hoyos et al., 1991), and are subject to inherent personality traits such as risk-seeking (Lehto, 1997).

Behavior modification programs often attempt to change behavior by manipulating the perceived probability of an outcome and its perceived value (Burkardt, 1981). That is, behavior modification attempts to move the subject's internal decision criteria toward a bias to choose behaviors that the psychologist desires. This is done by altering perceptions of likelihood for given outcomes through education and frequent reminders, thereby biasing the internal probability assessment. In addition, perceived values of selected behaviors are changed by increasing the positive value of a desired behavior and decreasing the value for following undesirable behavior (i.e., rewards and punishment, respectively). For example, the behavior of speeding along a highway can be reduced by either increasing the driver's expectations of getting caught (by increasing the number of patrol cars), or by significantly increasing the cost of getting caught (imposing large fines for infractions). Unfortunately, if the behavior reinforcements offered by behavior modification programs are not supported or sustained, the gains in health and safety flatten out and subsequently return to the baseline level (Komaki et al., 1978; Chhokar and Wallin, 1984). Cost/benefit models can help the human factors professional understand why workers bypass, disable, or otherwise compromise safeguarding systems. Once we understand the worker's perception of the costs and benefits of these unsafe behaviors, we can design systems that are compatible with these perceptions. As with physical ergonomics, this approach attempts to fit the work system to the natural human thought process, rather than attempting to change the human's decision-making process in favor of the desired behavior.

At this point, the author concedes that there exists little or no scientific research validating the application of the cost/benefit approach to the behaviors associated with compromising machine safeguarding. Clearly, this is fertile ground for safety research. The approach appears to have merit, however, since the design strategies suggested by it clearly reduce negative safeguarding behaviors.

Proceeding with the cost/benefit supposition, we need to identify the perceived costs and benefits of following alternative behaviors when working with machine safeguarding systems. Let's begin by defining two mutually exclusive behaviors:

A. Operate the equipment with the safeguarding system intact as designed.

B. Operate the equipment but compromise the safeguarding system.

This is obviously a simplification since workers can choose from among several behaviors, including refusing to operate the machine, or choosing the extent to which safeguards will be compromised. Further, it is assumed that compromising a safeguard is a deliberate act, and not a sudden reaction to a stressful situation.

Let's assume that the machine operator has to deal with a poorly designed safeguarding system (i.e., from the human factors perspective). Choosing A, safeguarding system intact, the worker expects the following benefits and costs.

Decision A Benefits:

- no injury (pain avoidance).
- supervisor approval if safety is a cultural value.

Decision A Costs:

- increased metabolic energy expenditure.
- increased time to perform the task and reduced productivity, with possible repercussions from the supervisor.
- increased mental workload and cognitive stress.

Choosing B, safeguarding system bypassed or compromised, the worker expects another set of benefits and costs.

Decision B Benefits:

- less energy expenditure.
- less operating time, resulting in more social time and possibly positive feedback from the supervisor recognizing productivity.
- excitement inherent with risk-taking.
- possible acceptance or even encouragement from peers.

Decision B Costs:

- risk of injury to both oneself and co-workers. (Workers seldom fully appreciate either the odds of injury or the potential severity.)
- discipline if caught compromising the safety system.
- mental stress associated with knowingly committing an unsafe act (conscience).

Assuming the cost/benefit analysis above accurately portrays the plant situation, it is easy to understand why the worker might choose alternative B. Behavioral programs using discipline obviously instill a cost for following strategy B, and the severity of discipline and the vigilance of the supervision determine the expected cost. Similarly, positive reward systems associated with "safety observations" provide a benefit for choosing behavior A. These management actions can obviously affect decision-making. Alternatively, human factor principles can also be applied to the safeguarding system design to achieve even better results, but without the ongoing costs of behavior maintenance. Ergonomics and human factors applied in design can reduce the expected costs of following strategy A, and may even negate the benefits of choosing behavior B. That is, good human factor design complements traditional behavioral approaches.

7.4 PRINCIPLES FOR HUMAN FACTORS IN SAFEGUARD DESIGN

Principle #1 – Design safeguarding to benefit the worker. Operator safety is an obvious safeguarding benefit. Good guarding design incorporates features that provide additional benefits. The following examples illustrate this principle.

1. Incorporate guides or flanges at guard openings to facilitate quick, accurate insertion of a workpiece, thereby increasing productivity.
2. Place tool rests strategically to reduce energy expenditure and grip force requirements.
3. Provide splash protection guards to decrease discomfort resulting from wet clothing.
4. Use mirrored surfaces on guards to improve visibility, by increasing light levels at the interest point, and by enhancing viewing angles.

Principle #2 – Design out the objectionable characteristics traditionally associated with machine safeguards. Negative characteristics can easily be identified by talking to workers in operations where safeguarding systems have been compromised. Examples of common complaints include:

1. Too hard to see through or around.
2. Too heavy to move.
3. Too big to reach around or over, and when reaching over, the guard rubs against me.
4. Too difficult to remove or replace.
5. Takes too much time to remove or replace.
6. Get tired of taking guard off every time machine needs lubrication.
7. In the way of jam clearing.
8. Too many accidental trips of the emergency stop or interlock system.
9. Sharp edges and protruding bolts keep catching clothing or causing cuts.

As exemplified in the next section, most of these complaints can be effectively addressed through careful design.

7.5 SELECTED HUMAN FACTORS TRICKS TO IMPROVE SAFEGUARDING

- Quick connections – Attach guards to machines using hinges, hangers, etc., that facilitate quick removal; and then secure the guard with a single fastener that requires a tool to remove, in order to maintain regulatory compliance.
- Emergency stop access – To prevent inadvertent trips of E-stop buttons, orient the button so that it is slapped downward to actuate, rather than across (i.e., mount on the top surface of an electrical box). Alternatively, mount the button above shoulder height; or place a protective shield along one side of the button so that a slap still actuates the switch, but rubbing against it generally will not.

- Visibility – Paint expanded metal guards traditional safety yellow or orange, but then paint a flat black "window" section. The human eye is naturally attracted to the bright orange or yellow and pays attention to it. Flat black or charcoal gray is minimally salient and the human naturally gives it less attention. The result is a guard with a section that is easy to see through. Alternatively, guards can be constructed of piano wire and made virtually invisible.
- Housekeeping – In areas where debris tends to accumulate behind guards, housekeeping can often be facilitated by installing dropouts, clean out trays, or deflector shields. In some cases, the guard itself may even dually serve as a deflection shield.
- Local disconnects – Equipment that needs to be locked out frequently needs to have a power disconnect located as close as possible to the location where a worker will need to work on the equipment. This minimizes the energy expenditure and time (costs) associated with achieving proper lockout.
- Interlock design – Self-checking and control-reliable interlock systems increase the cost associated with defeating interlocks. Interlocked gates, that will not permit opening until a machine has stopped, prevent operator errors and increase the cost of defeating them.
- Handle placement – Large guards often require handles to facilitate coupling for the worker during guard removal and replacement. Locate and orient handles so that lifting postures are as close to ideal as possible, paying special attention to shoulder and wrist postures.
- Guard size – Large guards should be hinged or otherwise attached so that lifting is not required to gain access.
- Fitts' Law and targeting – Guards are often left off not because they are difficult to remove, but because they are difficult to replace. Aligning pins or holes of guards with mating features is a task that follows the principles of Fitts' Law (Fitts, 1954). Adding flanges or funnels to fastening points increases the target-to-object ratio, thereby reducing the time, energy, strength, and stability required to orient a guard. Quick attachment characteristics should allow guards to be quickly set into position and supported by the machine while a single fastener is applied to achieve security.
- Simultaneous contact points – Guards should be designed such that replacement does not require simultaneous contact of two or more points. That is, a guard should be constructed so that only one pin or hole needs to be aligned at a time.
- Dual functions – Whenever possible, guards should perform multiple functions to increase their perceived value to the operator. For example, guards can often double as guides for part insertion, as noise enclosures, or as mirror panels.
- Weight and balance – Guards should be balanced about their handles or anticipated pick points. Minimize weight for any guard that is to be manually removed. Conversely, guard weight can be increased to prevent manual lifting; and lifting eyes can be provided to encourage the use of mechanical lifting equipment.

- Weight and targeting – Heavy guards are difficult to position accurately during installation, so supports and increased target-to-object ratios are needed.
- Presence-sensing – Interlocked presence-sensing systems are passive and do not require operator decision-making during stressful non-routine operations.
- Two-hand control – Two-hand controls often eliminate the need for additional point-of-operation guarding and allow unobstructed visibility. New no-touch or light-touch controls sense hand presence without increasing the risk of upper-extremity disorders associated with traditional palm buttons. Care should be taken during control locations to avoid non-neutral shoulder postures.

7.6 CONCLUSION

As with many aspects of safety, communications with operating and maintenance personnel are key to achieving successful safeguarding systems. By understanding the worker's complete job, including non-routine tasks, one can appreciate the safeguarding system attributes that are likely to encourage both safe and unsafe behaviors. Judicious application of human factors in design can effectively reduce the expected "costs" of using safeguards as intended, and may even negate the "benefits" of choosing to compromise them. The value of this approach is that it is passive, involves little or no ongoing cost, and is often more effective than behavior modification programs.

7.7 REFERENCES

Burkardt, F., 1981, *Information und Motivation zur Arbeitssicherheit* (Wiesbaden: Universum Verlagsanstalt). Cited by Zimolong, B., 1997, Occupational risk management. In *Handbook of Human Factors and Ergonomics*, 2nd ed., edited by Salvendy, G. (New York: John Wiley & Sons), pp. 989-1020.

Chhokar J., and Wallin, J., 1984, Improving safety through applied behavior analysis. *Journal of Safety Research*, **4**, pp. 168-178. Cited by Zimolong, B., 1997, Occupational risk management. In *Handbook of Human Factors and Ergonomics*, 2nd ed., edited by Salvendy, G. (New York: John Wiley & Sons), pp. 989-1020.

Fitts, P., 1954, The information capacity of the human motor system in controlling the amplitude of movement. *Journal of Experimental Psychology*, **47**, pp. 381-391.

Hoyos, C., et al., 1991, Vorhandenes und erwunschtes Wissen in Industriebetrieben. *Zeitschrift fur Arbeits- und Organisationspsychologie*, 9, pp. 68-76. Cited by Zimolong, B., 1997, Occupational risk management. In *Handbook of Human Factors and Ergonomics*, 2nd ed., edited by Salvendy, G. (New York: John Wiley & Sons), pp. 989-1020.

Johnson, L., 1999, The luxury of machine guarding. *Occupational Health & Safety*, **12**, pp. 26-29.

Komaki, J., et al., 1978, A behavioral approach to occupational safety: Pinpointing and reinforcing safe performance in a food manufacturing plant. *Journal of Applied Psychology*, **4**, pp. 434-445. Cited by Zimolong, B., 1997, Occupational risk management. In *Handbook of Human Factors and Ergonomics*, 2nd ed., edited by Salvendy, G. (New York: John Wiley & Sons), pp. 989-1020.

Lehto, M., 1997, Decision making. In *Handbook of Human Factors and Ergonomics*, 2nd ed., edited by Salvendy, G. (New York: John Wiley & Sons), pp. 1201-1248.

Nisbett, R. and Ross, L., 1980, *Human Inference: Strategies and Shortcomings of Social Judgment* (Englewood Cliffs, NJ: Prentice-Hall). Cited by Lehto, M., 1997, Decision Making. In *Handbook of Human Factors and Ergonomics*, 2nd ed., edited by Salvendy, G. (New York: John Wiley & Sons), pp. 1201-1248.

Wickens, C., 1984, *Engineering Psychology and Human Performance*, (Columbus, OH: Charles E. Merrill).

Zimolong, B., 1997, Occupational risk management. In *Handbook of Human Factors and Ergonomics*, 2nd ed., edited by Salvendy, G. (New York: John Wiley & Sons), pp. 989-1020.

Zimolong, B., 1985, Hazard perception and risk estimation in accident causation. *In Trends in Ergonomics/Human Factors II*, edited by Eberts, R. and Eberts, C. (Amsterdam: Elsevier). Cited by Zimolong, B., 1997, Occupational risk management. In *Handbook of Human Factors and Ergonomics*, 2nd ed., edited by Salvendy, G. (New York: John Wiley & Sons), pp. 989-1020.

CHAPTER EIGHT

Overcoming Traditional Job Rotation Obstacles
In the Design of a Manufacturing Cell

Richard Wyatt, Ph.D., PE, CPE, Chuck Davison,
Fletch Rickman and David Clampitt

8.1 ABSTRACT

Many manufacturing products are built and packaged in a traditional "build and package" process. Raw materials and other parts are brought into a series of workstations and assembled to form a saleable product. The product is then combined with needed accessories (i.e., instructions and other parts that are installed by the users) followed by packaging, labeling and shipping tasks.

This paper outlines two approaches to manufacturing cell design from a methods development standpoint, concentrating on the ergonomic and economic aspects of the design. This paper compares two separate and sequential cell designs that differ in job methods, with concentration on the economic and ergonomic effects of full job rotation within the cell.

8.2 INTRODUCTION

Many companies have improved their processes and lowered their manufacturing costs using demand flow, lean manufacturing, and Kaizen techniques. Actually, these techniques use a similar premise: continually removing waste from the process. Over a two-year period at Dana Corporation, set-up time was reportedly reduced 83 percent, while cycle time was reduced 85 percent (Cuscela, 1998). Reduced set-up and cycle times improve process flow, and the elimination of non-value-added work improves the financial bottom line. However, as each individual workstation becomes more efficient, idle time, which also provides worker recovery time, is eliminated. Job rotation becomes more and more important, especially when the production-line jobs are hand intensive.

Many benefits have been reported from job rotation (Jonssom, 1998; Hazzard et al, 1992; Henderson, 1992; MacLeod & Kennedy, 1993). These include:

- Reduced cumulative trauma disorders.
- Reduced job turnover.
- Increased innovation.

- Reduced work stress.
- Increased ability to handle change.
- Reduced turnover.
- Reduced absenteeism.

As pointed out in the same studies, implementing job rotation can be difficult. Many of the difficulties stem from system problems in the organization, employee perception and behavior, and not from the job rotation itself. The most common problems include:

- Difficulty in moving from one job to another.
- Machine operators not wanting to move or learn a new job.
- Differing wage formats.
- Employee training.
- Incorrect design or use of job rotation by management.

Job rotation allows workers to perform a variety of jobs, spreading the musculoskeletal risk of any one job task among all employees involved in the rotation. Mahone (1993) suggests that job rotation in itself would delay injury symptoms, and that engineering changes to the workstation should be considered first. Other studies suggest that a combination of engineering workstation changes be combined with job rotation to develop an effective injury control protocol (Helander, et al, 1993; McGlothlin, 1988; Frazier, 1978).

8.3 CELL DESIGN

Kaizen is a Japanese term meaning "continuous improvement" (Imai, 1986). The Kwikset Corporation, a leading manufacturer of door hardware, started using the Kaizen methodology in its Bristow, Oklahoma plant in 1997. Many processes were redesigned or moved closer together with the goal of decreased inventory, increased throughput, and reduced non-value-added material handling. The Bristow plant also used Kaizen techniques to redesign the final assembly processes – redesigning the linear assembly lines into smaller U cells. These U cells were designed using the principles outlined in Sekine (1992). These principles include basing the cycle time on customer requirements, centering production on assembly processes and changing the equipment layout to enable small-lot processing.

The plant's initial U cell was designed and balanced by a team of management and hourly employees who were very familiar with the assembly process. This exercise resulted in a cell requiring seven workstations: three were designated build stations, three packaging stations, and one material handling position.

With this U cell design, parts flow sequentially through the three build stations then through the three packaging stations. Builders take pre-made raw materials and assemble them into a usable product. In the packing stations, employees add necessary components (screws, instructions, and other field-required materials), place the components and the assembly into a box, combine completed units into a shipping box, and palletize the shipping boxes for shipment to the customer. The material handler's responsibility involves movement of needed materials to each of the workstations in a timely fashion.

The line balance for the cell is shown in the bar chart shown in Figure 8.1. Each of the critical work tasks was designed to eliminate as much motion from the job as possible. In designing the assembly process, the product was built completely before being combined with accessories and packed. Ergonomic workstation improvements were also implemented. For example, parts racks were moved closer to the employees, resulting in less reaching for both the builders and packers. Scissors lifts were added for the palletizing operation, resulting in less bending for that task.

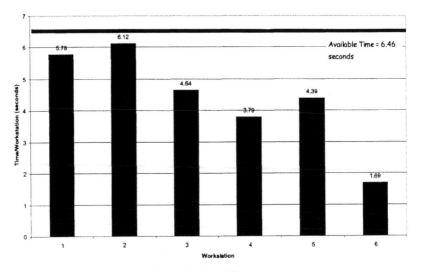

Figure 8.1 Line Balance.

Builders (workstations 1–3) have traditionally been separate job classifications when compared to packers (workstations 4–6). Therefore, in a given day, builders only build, and packers only pack. Part of the reason for the division is that the skill level required to become a proficient builder is far greater than that of a proficient packer. Within the U cell workstations, builders rotated hourly with the other two build stations, while the packers rotated hourly through the packing jobs. Should packers wish to learn the build process, a training process is available so they can develop the skill and speed necessary to become proficient at the job.

The line balance data in Figure 8.1 shows that the three most highly loaded workstations are the build stations. Build stations one and two pace the line; and their processing times of 5.78 and 6.12 seconds, respectively, are very close to the available cycle time of 6.46 seconds per piece. When the builders rotate, they move between heavily loaded workstations, with little rest and increased psychological stress. Packers rotate between stations that are not as heavily loaded, resulting in less ergonomic and psychological stress.

8.4 WHAT'S A SUPERVISOR TO DO?

Injury statistics for this cell have been compiled since its inception. During the calendar year 1999, two recordable repetitive-motion injuries have been logged in the OSHA–200 log. Both of these injuries occurred to employees involved in the build process. No injuries have been recorded for the packaging jobs.

An immediate response here might be quite apparent – rotate the builders to the packing jobs and vice versa. If this is done with a group of experienced employees that know all jobs in the cell, this rotation can work. But some employees do not like to build, others do not like packing, and others do not have the required coordination to build efficiently. Also, since the critical processes defining cell output are build processes, placing an inexperienced employee in the first build position will slow down the entire cell. Therefore, builders rotate with builders and packers rotate with packers.

8.5 MIXED BUILD-PACK DESIGN

In early 1999, another Kaizen team was established to redesign another assembly process. This team also included representatives from safety, engineering and production. One of the goals of this team was improved rotation. Based on lessons learned from the previous U cell, the design team decided to create one job class and eliminate the skills differential between builders and packers. The goals for the new cell design were:
- Eliminate employee injuries within the build and pack processes.
- Increase productivity 15%.
- Decrease manufacturing cycle time 95%.
- Decrease floor space requirements 50%.
- Increase production system flexibility (5 vs. 3 different products processing at the same time).

The new U cell is smaller, reducing travel distance and increasing employee-to-employee communication. Additional ergonomic design improvements have been made (i.e., improved parts presentation) and a completely different job rotation plan is used.

The new cell was implemented and compared to a sister line that runs the exact same product. Ergonomic and production statistics suggest several improvements, including reduced repetitive-motion risk, increased throughput and increased production. Six variables were used to judge the overall success of the cell and are shown in Figure 8.2. Since the cell start-up in March, 1999, no repetitive motion injuries have been recorded. This is compared with 5 injuries on the sister line. When comparing the number of pieces produced per operator per hour, the new cell output is 80.3 pieces/employee/hour, compared to 75.5 for the sister line. Work-in-process is reduced, due to decreased in-process inventory. The new cell also requires far less floor space than the older line. The new cell design has also reduced cycle time from 18 minutes to 1.5 minutes. All safety and production data obtained during the 12-week study suggests that this cell design is superior to the older line.

Measurement	Baseline (Sister Line)	Actual (New Cell)	Target Goal (New Cell)	Delta *
Repetitive-Motion Injuries	5	0	0	-5
Productivity (Pieces/Hour/Operator)	75.5	80.3	80	4.8
Cycle Time (Cell exit rate in minutes)	18	1.5	1	-16.5
Space Requirements (Square Feet)	600	295	500	-305
Work-In-Process (Number of Pieces with System Loaded)	200	15	5	-185
External Set-Up Time	20	5	3	-15

*Actual Minus Baseline

Figure 8.2 Cell Performance Metrics.

8.6 CONCLUSIONS

The above data suggest that overall productivity can be increased, while simultaneously improving the ergonomic well-being of the employee. The mixed build/pack job design did not and has not as of the date of this publication resulted in any repetitive-motion injuries. However, we cannot conclude that these results are the effect of the job rotation plan alone, since the job rotation plan was introduced simultaneously with engineering improvements designed to reduce bending and reaching. However, the build/pack job rotation schedule has proven to be effective in reducing the risk of cumulative trauma disorders, with no impact on production, when combined with engineering improvements and compared to the older production system design. Further analysis and additional Kaizen improvement will be performed on both generations of these cells. Similarly, the productivity and in-process inventory data are a result of twelve-week window and needs to be further analyzed over a longer time period.

8.7 REFERENCES

Cuscula, Kristin, April, 1998. Kaizen Blitz Attacks Work Processes at Dana Corporation, *IIE Solutions*, pp. 29-31.
Frazier, T.M., 1978. Job Satisfaction and Work Humanization: An Expanding Role for Ergonomics, *Ergonomics*, **21**(1), pp. 11-19.

Hazzard, L., Mautz, J., Wrightsman, D., 1992. Job Rotation Cuts Cumulative Trauma Cases. *Personnel Journal,* **71**(2) pp. 29-32.

Helander, M.G, Grossmith,E.J., Prabhu, P., 1992. Planning and Implementation of Microscope Work, *Applied Ergonomics,* **22**(1) pp. 36-42.

Henderson, C., 1992. Ergonomic Job Rotation in Poultry Processing, *Advances in Industrial Ergonomics and Safety, IV,* 443-450.

Jonsson, B., 1988. Electromyographic Studies of Job Rotation, *Scand J Work Environ Health,* **14**: supp. 1, pp. 108-109.

MacLeod, D., Kennedy, E., 1993. Job Rotation System: Report to XYZ Co. (On-Line, Available at website: http://209.85.88.134/job.htm).

McGlothlin, J.D., 1988. *An Ergonomics Program to Control Work-Related Cumulative Trauma Disorders of the Upper Extremities,* Dissertation, University of Michigan.

Sekine, Kenicihi, 1992. *One-Piece Flow Cell Design for Transforming the Production Process,* Productivity Press, Cambridge, Massachusetts.

CHAPTER NINE

High-Tech but Low-Cost Ergonomic Solutions

Dr. Joseph R. Davis, PE, CPE, CSP

9.1 ABSTRACT

This paper serves to show that high-tech ergonomic solutions can be low-cost and very effective as interventions to improve workplace ergonomics. This is illustrated by examining actual ergonomic interventions at three industrial companies in North Carolina. First, a small-sized (44 employees) printing company is examined to show low-cost usage of voice input/output technologies to reduce intensive hand keying typically associated with inputting news articles for publishing. Second, a medium-sized (350 workers) manufacturer of metal products is analyzed to show effective ergonomic solutions via high-tech materials (powder coatings) and high-tech equipment (robotic welding and automated spray painting) that eliminated ergonomic hazards while achieving production leaps by doubling output with a one-year payback. Third, a large-sized (2700 workers) poultry processing facility is examined to show that selective automation of ergonomically stressful tasks allowed the company to alleviate ergonomic stressors while saving $2 for every dollar invested. Thus, these case studies serve to illustrate that technology-based solutions can be highly effective for alleviating or eliminating ergonomic hazards, but such solutions can also still be low-cost. Additionally, it shows that ergonomists should not be content to stop with solutions that produce limited results. If only low-tech solutions are used, ergonomic improvements should be expected to stagnate eventually and reach a plateau. To improve beyond such a plateau, other approaches will become necessary. Therefore, as technological advances continue to occur and inevitably make new technologies easier to use, ergonomists should consider using high-tech but low-cost ergonomic solutions.

9.2 INTRODUCTION

History is rich with examples of technologies as both causes and cures for ergonomic maladies. As one example, the telegraph became widely used as a new technology for communication during the mid-1800's to early 1900's, and many people were employed as telegraph operators. Due to repetitious motions of the hand and wrist, many telegraph operators developed a condition known as "telegraphist's cramp" similar to today's carpal tunnel syndrome. A study of 8,153 telegraphists in 1911 showed that 64% of workers were experiencing some type of physical problem (Dembe, 1996, p.39). If the only interventions were low-tech (e.g., training, layouts, reaches, postures, job rotation, stretches), then little if any improvement would be expected because low-tech solutions would not solve the basic causal factor which was that the telegraph technology required precise and spasmodically repetitive motions of the hand and wrist.

Fortunately, "telegraphist's cramp" was eliminated via a progression of newer replacement technologies using the teletype (typewriter that sent Morse code) and then the telephone along with modern equipment such as fax machines. But this technological cause-and-cure cycle is now being repeated due to the increasing usage of the computer keyboard for communications via email, so new technologies (e.g. voice input/output or I/O) are emerging as solutions. Therefore, because history has shown us that technologies are often both culprits and cures for ergonomic maladies, it is reasonable for ergonomists to be continually seeking new high-tech replacement technologies to alleviate ergonomic problems.

The term high-tech generally refers to high usage of technology. Today, technology is a widely used word that is somewhat vague, as indicated by its definition in dictionaries as "the application of scientific knowledge, especially for industrial or business applications." Some experts on technological innovation provide a more precise definition of technology as "systematic knowledge made manifest by tools..." or "a system of components directly involved with acting on and/or changing an object from one state to another..." (Tornatzky et al, 1990, p.9). A related term is technological innovation, which can be defined as "situationally new development through which people extend their control over the environment..." (Tornatzky et al, 1990, p.xvi).

Perhaps the clearest understanding can be obtained by examining the roots of the word technology which originates from the Greek techne (tooled carving) and logos (logical thought or reasoning). Therefore, technologies can be considered as logically applied tools (hardware or software) by which we transform parts of our environment or change objects from one state to another. Most companies are endeavoring to achieve competitive success through cost reductions, quality improvements, and time-to-market advantages. For such companies, technological innovation can be defined as the introduction and usage of radically new tools or methods that achieve significant cost reductions, quality improvements, or leadtime reductions.

Embracing a new technology can cause a "leap-frog" situation (Porter, 1980, p.16), in which a company that previously was following industry leaders can quickly move from a position of being a follower to becoming a leader and win the competitive race via lower costs, better quality, and faster time-to-market. Well-known examples are the rapid "leap-frogging" that occurred when electronic calculators replaced slide rules, and when electronic digital watches replaced mechanical analog watches. Companies that had long been industry leaders quickly became losers. Also, ones that had previously been struggling as fledgling companies quickly became winners via high-tech strategies. Similarly, by utilizing high-tech solutions, companies that are lagging in ergonomic performance can quickly become leaders in their industry.

It is important for ergonomists to recognize that a natural maturation occurs for most technological innovations. Initially, a radically new technology will usually be costly and risky. As the technology matures, both the cost and the risk will decrease dramatically. Therefore, ergonomists should keep an open mind with regard to technologies (e.g., voice I/O and robotic automation presented herein), which at first consideration may have seemed too costly and risky, but then after technological maturation, these same technologies can clearly become viable as high-tech and low-cost solutions to improve workplace ergonomics.

A final introductory perspective is that ergonomists should not be content to stop with low-tech solutions that produce limited results. If an ergonomist notices a slowing rate of improvement from workplace ergonomic interventions, this is usually a signal that the ergonomist is overusing outdated technologies with diminishing returns on investments. As indicated by the historical example of the telegraph, if only low-tech solutions are used, then ergonomic improvements should be expected to stagnate eventually and reach a plateau. To improve beyond such a plateau, other approaches will become necessary. Therefore, as technological advances occur and continue to make high-tech solutions easier (lower cost and lower risk) to use, ergonomists should consider using high-tech solutions.

9.3 THREE CASES OF HIGH-TECH BUT LOW-COST SOLUTIONS

To show that high-tech ergonomic solutions can be low-cost and very effective for improving workplace ergonomics, actual ergonomic interventions were examined at three industrial companies in North Carolina. First, a small-sized (44 employees) printing company was examined to show low-cost usage of voice input/output technologies to reduce intensive hand keying typically associated with inputting news articles for publishing. Second, a medium-sized (350 workers) manufacturer of metal products was analyzed to show effective ergonomic solutions via high-tech materials (powder coatings) and high-tech equipment (robotic welding and automated spray painting) that eliminated ergonomic hazards while achieving production leaps by doubling output with a one-year payback. Third, a large-sized (2700 workers) poultry processing facility was examined to show that selective automation of ergonomically stressful tasks allowed the company to alleviate ergonomic stressors while saving $2 for every dollar invested.

9.4 CASE STUDY #1 (VOICE I/O FOR PUBLISHERS)

9.4.1 Background

The first case study of a high-tech but low-cost ergonomic solution was found in a small-sized (44 employees) printing company in North Carolina.

9.4.2 Problem

Employees were experiencing discomfort in their hands and wrists. This was clearly being caused by intensive hand keying which is typically associated with inputting news articles for publishing.

9.4.3 Analysis

To provide solutions for the company, an applied educational project was performed by North Carolina State University (NCSU). In this project, an ergonomics student worked under the direction and guidance of NCSU's Dr. Joe Davis (author), the specialist for industrial engineering and ergonomics within the Industrial Extension Service of NCSU's College of Engineering. The project involved workplace assessments via discomfort surveys, risk-factor checklists, videotaping work activities, and obtaining selected on-site measurements of forces, postures, and distances. These assessments identified at-risk workstations and tasks characterized by ergonomic stressors.

9.4.4 Alternatives Considered

The general finding was that some improvement could be achieved by low-tech solutions such as changing workstation layouts, modifying work methods, rotating workers, giving more rest breaks, and performing stretches. However, as typically occurs in many companies, the low-tech solutions provided only partial improvements.

9.4.5 Solutions Implemented

Because low-tech solutions had limited effectiveness, high-tech tools such as voice activated input/output (I/O) devices were recommended to reduce the intensive hand keying typically associated with publishing. Initially, voice I/O seemed too expensive and too radically new. As voice I/O technologies have matured, the cost has decreased by about a factor of 10 from 1997 prices of about $500 to the current prices of about $50. Also, voice I/O tools have become much easier to use as indicated by product marketing descriptions that now emphasize "naturally speaking" instead of the awkward stop-and-go dictation that was required only a few years ago. Not surprisingly, the previously "too radically new" technology of voice I/O is now commonplace.

9.4.6 Follow-up

The success of voice I/O as a high-tech but low-cost ergonomic solution was indicated during later follow-up in a testimonial from the company's Director of Finance who stated: "We appreciate receiving enlightening information about new technology tools such as voice-activated input devices." The company is continuing to explore and expand usage of voice I/O.

9.4.7 Critique

As documented in a report given to the company, by investing $1170 in the ergonomics project plus a few hundred dollars for voice I/O software, the company achieved at least an 11-to-1 payback on their investment. This payback was based on a savings of $13,920 by avoiding at least one cumulative trauma disorder (CTD) that would likely have otherwise occurred in the company's workforce. The $13,920 was the average cost from actual worker compensation cases for CTD's in North Carolina as summarized by the North Carolina Department of Labor. Hence, this is a good first example of a high-tech solution (voice I/O) that initially seemed too expensive and too radically new, but then as the technology matured, it was clearly shown to be optimal as a high-tech but low-cost solution for improving workplace ergonomics.

9.5 CASE STUDY #2 (ROBOTICS FOR METALWORKING)

9.5.1 Background

This second case study examines ergonomic interventions at an industrial workplace in North Carolina, where a medium-sized (350 employees) company manufactures metal products which are primarily metal cabinets for the telecommunications industry. This manufacturer utilizes a sequential production process that begins with raw steel components (sheet metal, angle iron, steel bars, etc.) which are cut and formed into shape, then assembled by welding together, and finally painted to create products such as metal cabinets that are used in the telecommunications industry to contain wiring and electronic circuit cards. This study shows effective ergonomic solutions via high-tech materials (powder coatings) and high-tech equipment (robotic welding and automated spray painting) that eliminated ergonomic hazards while achieving production leaps by doubling output with a one-year payback.

As with the first case study, an applied educational project was performed by having an ergonomics student work under the direction and guidance of NCSU's Dr. Joe Davis (author). The project involved workplace assessments via discomfort surveys, risk-factor checklists, videotaping work activities, and obtaining selected on-site measurements of forces, postures, and distances. This information was analyzed to identify at-risk workstations and tasks characterized by ergonomic stressors. This study is described in detail elsewhere (Davis, 1999a), so only the main points are summarized here.

9.5.2 Problem

The company was faced with the need to ensure an ergonomically safe and healthy workplace, while significantly increasing production output to satisfy rapidly growing customer demands for telecommunications cabinets as the company's primary product. This situation is often encountered by manufacturers who need to balance ergonomics along with stay-in-business requirements such as on-time right-quantity shipments to maintain customer satisfaction. Moreover, due to a strong economy with low unemployment (less than 1%) in North Carolina, it was impossible to find enough dependable skilled workers needed to reduce overtime requirements for existing workers in the primary production constraint areas which were welding and painting.

These welding and painting tasks were characterized by ergonomic stressors such as static high-force hand grasps (to hold welding tools or paint sprayers) and dynamic repetitive extreme reaches (above shoulders and below knees). Moreover, the work was characterized as "4-D" (dangerous, difficult, dirty, disappointing), which typically identify tasks that should be automated as described in detail within other writings (e.g., Davis, 1998). The first 2-D's, danger and difficulty, existed because of the ergonomic stressors. The second 2-D's, dirty and disappointing, were due to airborne contaminants from welding or paint spraying, and also because welding or painting all day is boring and unsatisfying work. Such 4-D jobs include tasks that people could do, but that are better for machines to do. Therefore, these are tasks that people should not do, and if people were to do these tasks, they would eventually wear out their bodies (backs, hands, wrists), be at greater risk of developing health problems (respiratory ailments), or become dissatisfied with their jobs and seek employment elsewhere. Thus, decisions were needed to improve ergonomics.

9.5.3 Analysis

The analysis regarding ergonomics is similar to most other business decisions, because the typical decision sequence used by company owners or designated decision makers is first to establish goals, next identify general strategies that include viable actions to achieve the goals, then evaluate how well each alternative would likely achieve the desired goals, and lastly select the action alternative that best achieves the goals. Strategies can be generally defined as "the means determined by the prevailing level of technological development," and action alternatives as the "strategies employed toward achieving given goals" (Zeleny, 1982, p.106). As explained within other writings (e.g., Davis, 1995a, 1995b, 1996, 1998, 1999b), the strategies and action alternatives for ergonomic decision situations can be summarized in the form of a decision hierarchy as shown in Figure 9.1.

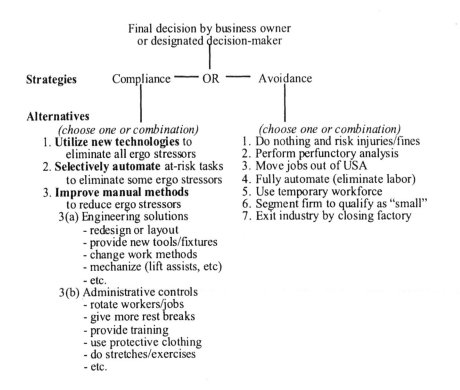

Figure 9.1. Decision hierarchy showing decision possibilities.

9.5.4 Alternatives Considered

Within each of the two general strategies of compliance or avoidance, there are action possibilities ranging from full compliance by eliminating ergonomic stressors to extreme avoidance by doing nothing while risking injuries and OSHA fines. In Figure 9.1, notice that compliance actions are in three basic categories which are: (1) utilizing new technology to eliminate *all* ergonomic stressors, (2) selectively automating at-risk tasks to eliminate *some* stressors, and (3) improving manual methods to *reduce* (but not eliminate) stressors via a variety of low-tech engineering solutions and administrative controls. For improving manual methods, ideas for engineering solutions and administrative controls can be found in publications such as *Elements of Ergonomics Programs* by NIOSH (1997). Also in Figure 9.1, notice that avoidance alternatives include doing nothing, performing perfunctory analysis, moving jobs out of the USA, fully automating production, using temporary workers, segmenting the company, and exiting the industry. Except for the extreme avoidance alternative (doing nothing), the avoidance actions actually constitute legal compliance because they do satisfy the "letter of the law," but they do not satisfy the "spirit of the law," so they are categorized here as avoidance strategies. The avoidance alternatives are not advocated by the author, but it is necessary to include these for completeness of the decision hierarchy.

9.5.5 Solutions Implemented

Initially for the company in this case study, quick-fix ergonomic improvements were accomplished via low-tech solutions such as job rotation and increasing the density of manually placed parts while allowing workers to take more rest breaks at their discretion. Through such interventions, modest improvements for ergonomics (less discomfort) and productivity (20% output increase) were achieved, but improvements for both ergonomics and productivity reached a plateau when only low-tech solutions were used. Thus, more effective solutions were implemented by using high-tech materials (powder coatings instead of solvent-based paints) and high-tech production equipment (robotic welding and automated spray painting). Through these high-tech solutions of robotic welding and automated spraying of powder paints, ergonomic hazards were eliminated (no discomfort), while simultaneously achieving leaps (doubled output) in production with cost payback in approximately one year.

Automation is defined to occur when an intelligent mechanical system provides both movement and control (people are involved only for maintenance and repairs) to accomplish a work task (Kroemer, 1997). Mechanization is a primitive type of automation that occurs when a mechanical system provides movement force but does not provide control (people are involved to provide control). A robot is an advanced type of automation. A robot is defined as a multi-functional manipulator designed to move materials, parts, tools, or special devices through variable programmed motions for the performance of a variety of tasks (Odrey, 1992, p.7.171). Key features are that robots operate automatically, have versatile operations, are computer controlled, are teachable, have multiple degrees of freedom, and react based on feedback from the environment. Not all industrial tasks can be automated, so it is clear that human labor will continue to be very necessary and valuable for jobs involving variable-path decision-intensive tasks, especially if work activities are occasionally based on poorly-defined vague information that requires judgment for proper interpretation. However, there are specific ergonomic hazards that are particularly suited for automated solutions, in lieu of having people perform manual work.

In particular, tasks that involve high-speed fixed-path repetitive assembly motions are likely first candidates for selective automation. Due to the high-speed nature of the work, an automated solution with control provided by a system is preferable over a mechanized solution with control provided by a person.

For high-speed fixed-path hazardous work tasks, the best solutions are by way of: 1) automated assembly systems ("hard" automation) for high volume production; 2) progressive assembly indexing tables with PLC controlled ("soft" automation) linear actuators for moderate volume production; or 3) flexible robots ("agile" automation) for mass-customized production.   With increasing emphasis on producing large volumes comprised of customized small production lots (termed "mass customization"), flexible robots have been increasing in popularity.

As evidence of the increasing industrial emphasis on robotic solutions for mass customization, sales of industrial robots have increased over 172% during a recent five-year period (Wells, 1998), and the key factor driving this growth is the need to produce mass-customized products with improved quality and with reduced direct labor costs (Hill, 1995; Wells, 1998). Many suppliers now offer robotic equipment as modular building-block subsystems (base unit, movement arms, proximity sensors, end effectors, parts feeders, etc.) that plug together to create a functional system. The typical cost of hardware and software for a multi-axis programmable robot is approximately $50,000 for initial purchase, plus about 10% per year for maintenance. More precise and up-to-date information concerning suppliers and costs can be found via internet searches using keywords such as "automation," "robot," and "automatic painting."

Automation is no longer the exotic technology that it once was. Automation and robotics are technologies that are in transition from having been very high-tech and risky, but now they are becoming more commonplace and easy to implement. As technological maturation occurs, a strategy such as automation can become easier to implement, and hence become more likely to be chosen as the best action alternative to alleviate an ergonomic hazard. Within OSHA ergonomic guidelines (e.g., OSHA,1990; NIOSH,1997), automation is mentioned as an engineering solution that can be implemented after ergonomic hazards have been identified. Once hazards have been identified, it is prudent to eliminate or alleviate hazards as quickly as possible. Hence, rapid-deployment solutions are desirable.

Today, rapidly-deployed and cost-effective automation solutions are available either as off-the-shelf complete systems or as modular building-block subsystems. Thus, this case study serves to illustrate that for some (but not all) situations, selective automation can be the best ergonomic intervention to minimize costs, maximize quality, and eliminate ergonomic hazards quickly and simultaneously (Davis, 1995a, 1995b, 1996, 1998, 1999b). As a brief example of the viability and popularity of automated solutions to alleviate ergonomic hazards associated with high-speed fixed-path repetitive tasks. The following is an excerpt from a General Accounting Office (GAO) report that examined how ergonomics programs were implemented in five US facilities.

"By automating many of the steps in circuit board assembly over the last decade, Lewisville has eliminated much of the manual assembly work and, thereby, the associated ergonomic risks. For example, a stainless steel stencil is now laser-etched onto the board, an automated squeegee applies the paste to the board, and the boards are then fed into a machine that loads components via feeder reels and chip shooters. In these highly automated work areas, there are few ergonomic hazards." (GAO, 1997, p.126)

Advocating automation should not be misinterpreted as an anti-labor sentiment. Instead, it is simply asserting that humans should not be used as mindless mechanical manipulators for performing high-speed repetitive motions over fixed movement paths in dangerous, difficult, dirty, or disappointing work conditions.

These types of high-speed repetitive tasks are the primary culprits for musculoskeletal problems, especially when tasks involve high repetitions in combination with extreme forces, awkward postures, or rapid ballistic movements. Therefore, these are tasks that people should not do, and if people were to do these tasks, then they would eventually wear out their bodies (hands, wrists, etc.) or become dissatisfied with their jobs and quit. Hence, in selected situations, automation is truly more of a humanitarian, not anti-labor, solution as a high-tech ergonomic intervention to improve the workplace.

9.5.6 Follow-up

A one year follow-up showed that ergonomic interventions were effective in alleviating hazards because there were no new CTDs, and workers seemed more satisfied with their work conditions as evidenced by increased worker retention, while the company achieved a more than doubling of production output with a one year payback. This indicates a low-cost solution because "cost" is defined as the net present value of expenses (positive costs) combined with the net present value of savings (negative costs), So a one year payback indicates that the high-tech solutions via robotics and automation were actually low-cost because a net savings (cost less than zero, so clearly low-cost) occurred within one year. This also shows that ergonomic solutions can require initially high investments but still be considered low-cost.

9.5.7 Critique

This study serves to illustrate that high-tech materials (powder paints) and high-tech equipment (robotics) can be effective as low-cost ergonomic solutions. Additionally, it shows that ergonomists should not be content to stop with solutions that produce limited results. By using only low-tech solutions, such as job rotation and increased rest breaks, ergonomic improvements stagnated and reached a plateau for the company in this case study. To improve beyond that plateau, other approaches were needed. Therefore, high-tech but low-cost solutions were successfully implemented to improve workplace ergonomics. Hence, whenever a stagnating rate of ergonomic improvement via low-tech solutions is encountered, ergonomists should consider using high-tech solutions.

9.6. CASE STUDY #3 (AUTOMATION FOR POULTRY PROCESSING)

9.6.1 Background

An excellent example of ergonomic and economic win-win success via high-tech automation was found in the first major ergonomic settlement agreement by OSHA in North Carolina. OSHA actively provided information on this case to the public because "the Perdue investigations were the first major OSHA inspections in North

Carolina concerning cumulative trauma disorders, and the settlement agreements could represent a precedent for other employers with similar problems" (N.C.OSHA, 1991, p.v). This settlement agreement was first published by N.C.OSHA in a 187-page report (N.C.OSHA, 1991) titled "North Carolina OSHA Compliance Special Report - Perdue Farms." Ergonomic actions by Perdue were subsequently published during later years in North Carolina newspapers (N&O, 7/20/93, 8/5/93, 9/26/93, 1/16/95), and in national publications such as the Bureau of National Affairs (BNA, 1996). The following case study provides a summary of the Perdue Farms ergonomic saga.

9.6.2 Problem

Perdue Farms Inc. was the first company in North Carolina fined by OSHA for exposing workers to repetitive motion injuries (BNA, 1996). This was a major ergonomics case in North Carolina because it involved 2700 workers at Perdue's poultry processing facilities located in Lewiston-Woodville, NC and in Roberson, NC. Prior to the OSHA citations, there had been hundreds of complaints by injured employees. Perdue was cited for causing cumulative trauma disorders. The citations included 1 willful violation, 15 serious violations, and 5 non-serious violations (N.C.OSHA, 1991, p.5-15). Even with such a difficult start, Perdue has now become a showcase of ergonomic success and an example for other companies to follow. OSHA now highly praises the company for its cooperation (BNA, 1996).

9.6.3 Analysis & Alternatives Considered

Perdue's success was due in part to low-tech interventions, such as educating workers, instituting job rotation, and increasing rest breaks. However, as explained in the following paragraphs, ergonomic success was primarily due to automation as a high-tech solution.

9.6.4 Solutions Implemented

As explained in other writings (Davis, 1998), the first high-tech implementation project should make a "winner" out of everyone involved. Therefore, it is wise to choose the first project as the one that is the most feasible technically and thus has the greatest probability of successful implementation. Perdue did exactly that as illustrated by the following quote from a local newspaper.

Evisceration was one of the first places Perdue went in its effort to improve conditions... In evisceration, what used to wear out women's wrists and shoulders is now done by a robotic claw...One or two gloved workers stand by... if the machine misses anything... (N&O, 9/26/93, p.1c).

After excellent success from the first project, Perdue widely implemented automation solutions. To accomplish extensive automation successfully, design for automation (DFA) principles must be utilized as explained elsewhere (Davis, 1998). Applying DFA to poultry (floppy chickens) might seem impossible, but Perdue did that primarily by standardizing the size of incoming chickens to an average live weight of 4.85 pounds which was reached about 49 days after egg hatching. Thus, by embracing DFA principles, Perdue successfully automated, as indicated by the following examples.

If you had been in this plant just 3 or 4 years ago, you wouldn't even recognize it now... Feathers plucked by machines... Stainless steel production line... A machine makes a slice.. Another machine reaches... (N&O, 9/26/93, p.1c).

All of this used to be done by hand... Now the idea is to have machines do the repetitive work as much as possible... (N&O, 1/16/95, p.3a).

Much of Perdue's efforts have centered on using automation to take care of the jobs that hurt workers' hands and arms the most. Workers mostly play supporting roles to machinery as chickens are defeathered, eviscerated, inspected, cooled, cut, and wrapped... (Stancill, 10/9/95, p.5a).

9.6.5 Follow-up

By utilizing automation as a high-tech ergonomic solution, Perdue improved ergonomics while achieving economic success and competitive advantage as shown by the following excerpts from news articles.

Perdue's spending on workers comp went down from $3.6 million to $1.3 million... (Stancill, 10/9/95, p.5a)

Perdue's lost-time ratio is now one-eighth of the poultry processing industry... (N&O, 1/16/95, p.3a).

Perdue is saving $2 for every dollar invested in its ergonomics program. (BNA, 1996, p.10).

This 2-for-1 payback indicates a low-cost solution because "cost" is defined as the net present value of expenses (positive costs) and savings (negative costs). Therefore, a 2-for-1 payback indicates that the high-tech solution via automation was actually low-cost because it provided a net economic savings (cost less than zero, so clearly low-cost). Also, this shows that ergonomic solutions can require initially high investments but still be considered low-cost.

9.6.6 Critique

Perdue's success was due in part to low-tech interventions such as educating workers, instituting job rotation, and increasing rest breaks. However, ergonomic success was primarily due to automation as a high-tech solution. It is clear from this case study and other studies (e.g., Davis, 1995a, 1995b) that significant ergonomic and economic win-win success can be achieved via selective automation of hazardous tasks.

9.7 CONCLUSION

This paper serves to show that high-tech ergonomic solutions can be low-cost and very effective as interventions to improve workplace ergonomics. This has been illustrated by examining three cases of ergonomic interventions at companies in North Carolina. These case studies serve to validate the conclusion that technology-based solutions can be highly effective for alleviating or eliminating ergonomic hazards, and that such solutions can still be low-cost.

Another conclusion is that ergonomists should recognize and appreciate the natural maturation which occurs for most technological innovations. Initially, a radically new technology will usually be costly and risky. As the technology matures, both the cost and the risk will decrease dramatically. Therefore, ergonomists should keep an open mind with regard to emerging technologies (e.g., voice I/O and robotic automation presented here), which at first consideration may have seemed too costly and risky, but then after technological maturation, these same technologies can clearly become viable as high-tech low-cost solutions to improve workplace ergonomics.

A final conclusion is that ergonomists should not be content to stop with low-tech solutions that produce limited results. If an ergonomist notices a slowing rate of improvement from workplace ergonomic interventions, this is usually a signal that the ergonomist is overusing an outdated technology with diminishing return on investment. If only low-tech solutions are used, ergonomic improvements should be expected to stagnate eventually and reach a plateau. To improve beyond such a plateau, other approaches become necessary. Therefore, as technological advances occur and continue to make high-tech solutions easier (low-cost and low-risk) to use, ergonomists should consider using such solutions.

9.8 ACKNOWLEDGEMENTS

In this case study, technology assistance was provided by the North Carolina Manufacturing Extension Partnership (MEP) of North Carolina State University's Industrial Extension Service. As part of a national MEP network affiliated with the National Institute of Standards and Technology, NC MEP's mission is to help manufacturers improve competitiveness by adopting modern technologies and production techniques. There are MEP offices in every state of the nation to provide technology assistance, and details such as local offices in each state can be found via the internet at http://www.mep.nist.gov.

9.9 REFERENCES

BNA, 1996. Ergonomics success story - Perdue Farms. In *BNA Communicator*, Spring 1996 issue, (Rockville, MD: Bureau of National Affairs, BNA Communications Inc.), p. 10.

Davis, J., 1999a. Proactive ergonomics via technology-based solutions. From presentation at March 9-11, 1999 *2nd Annual Applied Ergonomics Conference* in Houston, TX. Published in *Applied Ergonomics Case Studies Volume 2*, edited by Alexander, D., (Norcross, GA: Engineering and Management Press), pp. 102-114.

Davis, J., 1999b. Ergonomics at a medical device assembly plant. From presentation at March 10-11, 1998 *1st Annual Applied Ergonomics Conference* in Atlanta, GA. Published in *Applied Ergonomics Case Studies Volume 1*, edited by Alexander, D., (Norcross, GA: Engineering and Management Press), pp. 119-131.

Davis, J., 1998. Practical ergonomic solutions via cost-effective rapidly-deployed automation. In *Ergonomic Process Management: A Blueprint for Quality and Compliance*, edited by Kohn, J., (Boca Raton, FL: CRC/Lewis Publishers), pp. 232-269.

Davis, J., 1996. A multi-criteria decision model for prescribing optimal ergonomic action strategies for industry. In *Human Interaction with Complex Systems: Conceptual Principles and Design Practice*, edited by Ntuen, C., (Boston, MA: Kluwer Academic Publishers), pp. 165 - 183.

Davis, J., 1995a. Automation and other strategies for OSHA ergonomics compliance. In *Industrial Engineering*, February 1995, **27**(2), pp. 48-51.

Davis, J., 1995b. Automation and other strategies for compliance with OSHA ergonomics. In *Proceedings of 1995 International Industrial Engineering Conference*, (Norcross, GA: Institute of Industrial Engineers), pp. 592-599.

Dembe, A., 1996. *Occupation and Disease*, (London: Yale University Press).

GAO, 1997. *Worker Protection: Private Sector Ergonomics Programs Yield Positive Results*, August 27, 1997 report of U.S. Congress's General Accounting Office (GAO) 140-page document number GAO/HEHS-97-163.

Hill, S., 1995. Robots are back!. In *Manufacturing Systems*, March 1995 issue, pp. 42-66.

Kroemer, K., 1997. *Ergonomic Design of Material Handling Systems*, (New York: Lewis Publishers).

N.C.OSHA, 1991. *North Carolina OSHA Compliance Special Report - Perdue Farms*, 187 pages, (Raleigh,NC: NC Department of Labor).

NIOSH, 1997. *Elements of Ergonomics Programs*, (Cincinnati, OH: National Institute for Occupational Safety and Health, NIOSH), 133-page document number DHHS/NIOSH-97-117.

N&O, 7/20/93, 8/5/93, 9/26/93, 1/16/95, Coping with pain... Perdue Farms. In *News & Observer* (N&O) newspaper articles by Shiffer,7/20/93,p. 1b; Quillin,8/5/93,p.3a; Quillin,9/26/93, p.1c; Press,1/16/95,p.3a; Raleigh, NC.

Odrey, N., 1992. Robotics and automation. In *Maynard's Industrial Engineering Handbook*, 4th edition, chapter 10 in section 7, edited by Hodgson,W., (New York: McGraw-Hill).

OSHA, 1990. *Ergonomics Program Management Guidelines for Meatpacking Plants*, Occupational Safety & Health Association (OSHA) document 3121.

Stancill, 1995. Perdue cuts injury costs with training, automation. In *Charlotte Observer* newspaper article on October 9, 1995, p.1a-5a; Charlotte, NC.

Porter,M., 1980. *Competitive Strategy: Techniques for Analyzing Industries and Competitors,* (New York: Macmillan Publishing).

Tornatzky, L., *et al*, 1990. *The Processes of Technological Innovation*, (Toronto: Lexington Books).

Wells, J., 1998. Robot sales boom as integration gets easier. In *Managing Automation*, August issue, part 2, pp. 4-10.

Zeleny, M., 1992. *Multiple Criteria Decision Making*, (NY: McGraw-Hill).

CHAPTER TEN

From Handtrucks to Palletizers: Advances in Material Handling Equipment for the New Millennium

George Chuckrow, CPE, CSP

10.1 A BRIEF HISTORY

Material handling and its related injuries have been with us ever since the first human crawled out of the sea and began to walk on dry land; or, if you prefer, since Adam and Eve.

Early attempts at solving material handling problems did not do much to reduce the physical load on the material handlers until the wheel was invented.

As society evolved, material handling evolved along with it. By the time the pyramids were built, various tools had been developed to transfer loads and their resultant stressors from people to hoists and other devices still in use today.

The handtruck, forklift, crane, and conveyor became the backbone of material handling. Then and now, they have moved materials efficiently and effectively, while reducing the load borne by the worker.

10.2 TODAY AND LOOKING FORWARD

Today's workers have been the beneficiaries of technological advances not even imagined at the end of the last millennium. These have come at a cost, though: a dramatic increase in work-related musculoskeletal disorders (WRMSDs).

Industry has developed production methods that allow us to manufacture more, at faster rates, and to do so with fewer employees, increasing the load on the workers who are left to do the jobs.

This is particularly apparent in material handling. Employers, who have been quick to embrace state of the art production techniques, have been less willing to modernize material handling, which is often still done the old way – by hand.

10.3 TAKING THE MANUAL OUT OF MATERIAL HANDLING

The technology that has allowed us to produce goods more rapidly and efficiently than ever before, has in recent years made its way to material handling.

The greatest strides have been made in the last 10 - 15 years, thanks largely to a growing awareness of ergonomics in industry, and a continuing rise in the incidence of WRMSDs, with a growing emphasis on ergonomic solutions to prevent them.

Thankfully, these strides do not take workers out of the production chain, but they do reduce the loads that are handled, along with the resulting physical stressors.

Several improvements on the traditional handtruck now exist; these include handtrucks with powered platforms to raise and lower materials, handtrucks with electrically powered drive wheels, and powered stair climbing handtrucks. Even when it is not possible to use a powered handtruck, traditional handtrucks have been improved upon in recent years. They are being built of lighter weight materials, with better handles and balance points.

Similar to handtrucks are wheeled lifts, capable of raising and lowering a variety of materials so employees do not have to lift or lower heavy loads manually, especially while bending at the waist. These lifting devices can be equipped with a variety of attachments that allow them to be used with anything from case goods, to large rolls of material like carpeting, to barrels and beer kegs. For load handling, they are usually electrically powered, but can be equipped with hydraulics if necessary; the simplest of these devices use a hand crank to raise and lower the load bed.

Palletizing and de-palletizing materials, together with stacking and unstacking directly on the floor, constitute one of the most common tasks in production facilities, and one that results in many back and trunk injuries. While automatic palletizers can eliminate these injuries, the price, both in terms of capital investment, and the loss of jobs they replace, is usually unacceptable.

A highly effective yet underused alternative to an automatic palletizer, is one of a variety of lift tables, often equipped with turntables, that allow materials to be handled at waist level, and directly in front of the employee, with no bending or leaning over the pallet. These can be completely self-contained and can automatically lift or lower, based on the amount of weight placed on them, by using springs or air cylinders, and can be moved to different work areas as needed (Figure 10.1). Electrically-operated and hydraulic lift tables can be installed at floor level or can be counter-sunk in floors to allow high stacking without requiring employees to lift above the shoulders. Tilt/lift tables (Figure 10.2) let workers maintain neutral back and upper extremity postures while lifting and reaching into large boxes and wire or metal containers. Mobile lift tables usually handle loads from about 500 pounds to more than 1,000 pounds, can be used to transfer materials to and from tables and shelves set at various heights, and can also be used as height-adjustable work stations (Figure 10.3).

Positioners or manipulators are articulated arms, that use a vacuum or gripping device to lift and lower materials. These can be permanently located, mounted on overhead rails (like a crane), or on wheels for use in various locations. The employee remains in the production loop with these, but the manipulator supports the load being handled, relieving stress on the worker's body. When combined with lift tables and conveyors, positioners and manipulators allow neutral postures throughout a range of material handling tasks, with a minimum of force exerted by, or imposed on, the employee. As an added bonus, they can be used to lift, rotate, and empty sacks, barrels, and other containers into production machines.

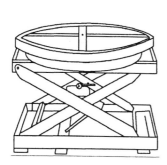

Figure 10.1 Lift Table.

Figure 10.2 Tilt/Lift Table.

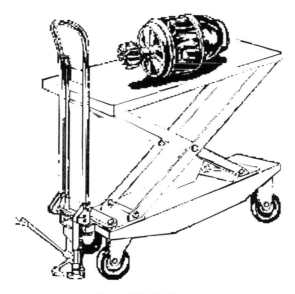

Figure 10.3 Mobile Lift Table.

Other machines that lift and dump include a range of barrel and drum handlers, along with a range of industrial lift/dump machines used to transfer raw materials from containers to production machines. Barrel and drum handlers are used to transport these items around a facility while imposing minimal force on workers. They are also used to invert and empty barrels and drums, eliminating the need for employees to do this manually. These machines can also be used for handling beer kegs, especially for loading them into car trunks. Lift/dump machines are typically used in food processing and other production facilities to raise tubs containing large quantities of raw materials, then empty them into a processing machine.

Several powered pushing devices have entered the market in recent years; these are used by employees to push anything on wheels, in any setting. Some grocery chains are using these to reduce the risk of injury associated with rounding up shopping carts in the parking lot. They are used to push "trains" of shopping carts back to the store by utility clerks (Figure 10.4).

A relatively simple device, that helps prevent back injuries when workers remove laundry or other material from carts known as laundry trucks, is the platform lift (Figure 10.5). This is usually a piece of plywood or a plastic sheet that fits just inside the inner dimensions of the cart, which is suspended by a hook at each corner that is attached to a spring. The platform rises and falls automatically as it is loaded. This prevents bending and reaching into deep carts, and helps prevent back and trunk injuries.

Figure 10.4 Powered Pushing Devices.   **Figure 10.5** Platform Lift.

One type of material handling device often overlooked is the conveyor. These have been around for years, but are not always designed with the employee in mind. Employees must frequently intervene in the conveying process, transferring materials from one conveyor to another, or from conveyors to different material handling devices, such as carts. A well designed conveying system uses a variety of horizontal and vertical conveyors, as well as various diverters, pushers, and drive devices (e.g., roll casings, ball transfers, and belts) to minimize handling.

Like conveyors, carousels—not the kind with the pretty horses—but those designed to move materials, are often overlooked. Vertical carousels (Figure 10.6) especially can be used to pick and place stock, eliminating awkward postures and high forces.

Forklifts and stockpickers have come a long way; narrow aisle lifts are now available for high density storage areas where standard forklifts cannot be maneuvered. A relatively new innovation in stockpickers is one which allows the employee's work platform to be raised or lowered independently of raising and lowering the forks that hold the pallet that items are stacked on; the forks can be set at one height, with the employee's platform set at a different height, allowing the employee to pick items while maintaining neutral back postures, instead of having to bend while cubing out the pallet. For those tasks where a forklift is too large and a pallet jack inadequate, stackers can be used. They're smaller, and less expensive than forklifts, but unlike a pallet jack, have forks that can be raised and lowered like a forklift's.

Figure 10.6 Vertical Carousel Conveyor.

10.4 MAKING THE INVESTMENT

Of course, most ergonomic solutions require some sort of investment, either in time, money, or both.

The investment in ergonomic solutions pays for itself by preventing injuries, and reducing the resulting workers' compensation claim costs, as well as administrative costs paid for right out of the uninsured employer's pocket. The reduction in claim costs ultimately results in premium reductions, either through experience rating or premium credits. When lost time injuries are prevented, overtime and other related costs are reduced.

Ergonomic solutions usually help optimize production and quality. This is just as true with a material handling device like a lift table, as it is for an automated packaging machine.

The actual amount of investment is going to depend on the type of solution applied. Lift carts start at less than $1,000, while lift tables usually range from $2,000 to $5,000. Manipulators, positioners, and vertical carousels start in the $10,000 range, but platform lifts cost less than $100.

Ultimately, the amount of capital investment necessary will depend on what type of equipment is being purchased, who the vendor is, and the number of units purchased.

Often, vendors will agree on lower unit prices if a purchaser commits to a certain volume, even if the purchase is spread over several years.

10.5 EQUIPMENT RESOURCES

Where does one find a positioner, lift table, or vertical carousel? To determine what equipment might be right for your business, and where to find it, you may need the services of a qualified ergonomist.

A good Certified Professional Ergonomist (CPE) will not only be up-to-date on what kind of equipment is available, but will also know where to get it.

Even if you know what it is that your business needs, the perspective of a Certified Professional Ergonomist may be of value.

If you choose to go it alone, subscribe to one of the many publications that are offered at no cost. *Material Handling Product News*, *New Equipment Digest*, and *Industrial Product Bulletin* are just a few that provide product information.

Other publications are also available. Perhaps the best resources are on the internet.

Penton Publications and Cahners Business Publications are the primary publishers of information digests for everything from manufacturing to bakeries to hospitals, with everything in between thrown in for good measure.

10.6 WHAT THE FUTURE HOLDS

As employers continue to seek solutions to their ergonomics-related problems, manufacturers will continue trying to build a better mousetrap. The key solutions to today's material handling problems have been reviewed here, and there are many other solutions already available. New material handling products are constantly being developed. As the demand increases, competition among material handling equipment makers increases as well, and prices drop.

Our aging work force is not going away, and advancements in medical science, along with changes in the social security and retirement systems are keeping people working longer than ever before. The need to provide workers with the tools to keep them unhurt and at work, will require new solutions to current and future challenges.

10.7 INTERNET SOURCES OF INFORMATION

Cahners Business Information Publications
8773 Ridgline Blvd.
Highland Springs, CO 80216
www.cahners.com
    Food Engineering — www.foodexplorer.com
    Industrial Product Bulletin — www.iph.com
    Material Handling Product News — www.mhpn.com
    Material Handling News — www.manufacturing.net/magazine/mmh
    Packaging Digest — www.packagingdigest.com
    Warehousing Management — www.warehousemag.com
Penton Publications
1100 Superior Ave.
Cleveland, OH 44114
www.penton.com
    Material Handling Engineering — www.mhesource.com
    New Equipment Digest — www.newequipment.com

CHAPTER ELEVEN

Development & Implementation for Two Specialized Material Handling Devices: Metal Plate Movement & Oven Loading for Precision Optics

Jeffrey D. Brewer

11.1 INTRODUCTION

The following two case studies describe design improvements that have reduced ergonomic stresses on employees in manufacturing environments. Case 1 involved a forceful pushing task commonly encountered by a technician in the saw shop of a machine shop. Case 2 dealt with frequent strenuous lifting, and precise manual manipulation above shoulder height of 35 pound canisters filled with glass optical elements. In both cases, a team-based approach was used to identify the problem, generate potential solutions, evaluate, test and implement the most desirable solution. In Case 1, the solution involved the use of in-house resources to create a "low cost" device for carrying out the forceful pushing. In Case 2, the solution involved the in-house creation of a more elaborate and expensive device to eliminate manual overhead loading and unloading.

Project successes were due to a cooperative effort between operators, managers, engineers, tooling personnel, site safety councils and a regional ergonomist. Success was indicated by task performance improvement and injury risk factor reduction. In both cases, ergonomic problems had been recognized, but intervention efforts had not resulted in solutions. Technical challenges and the perceived "high cost" of solutions were the primary reasons for difficulty in eliminating these strenuous activities.

11.2 CASE 1 - PROBLEM DISCOVERY AND QUANTIFICATION

Primary activities of the saw shop technician involve ordering materials to keep the shelves stocked, moving incoming metal to the shelf locations, removing requested metal (often cutting material to size), and delivering material to the internal shop customer. The majority of metal stock items are received in small lot sizes and can be transported from delivery zones by pallet jack and positioned on shelves either by hand or with the overhead gantry-type hoist positioned above the saw shop. One task that had proven to be a challenge for the technician involved pushing a four-foot by four-foot by two-inch aluminum plate weighing 449 pounds out of a storage rack. The technician accomplished this task by pushing with a steel rod braced against his leg near hip level (see Figures 11.1–11.3). With each exertion the plate would "inch" out of the rack. Lack of visibility during the push required the technician to walk around the rack one or more times to gauge plate displacement. This movement caused the pushing cycle time to vary between one to two minutes. The technician had received ergonomics training and was using a good manual pushing method. Experiments measuring the coefficient of friction showed that the manual pushing force required to move the plate exceeded 229 pounds. Generating this force required tremendous exertion by the technician.

Immediate ergonomic and safety concerns included the strain potential of pushing, and injuries related to a slip or fall, such as a sprain, cut or contusion. Productivity concerns included fatigue and the time required for the pushing task. Previous analyses by two other ergonomics specialists had determined that a larger saw shop was needed to allow plates to be stored flat in piles and that a vacuum lift should be installed. The current facility could not readily accommodate the increase in space. Coupling this with the $10,000 expense for an overhead vacuum system along with installation and end effector costs, led these solutions to be deemed infeasible by shop management.

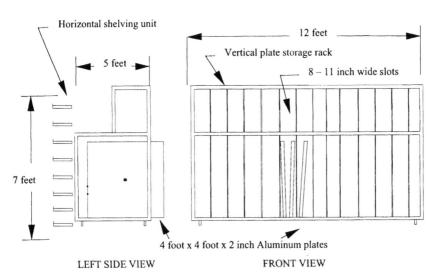

Figure 11.1 Side and front views of metal plate storage rack.

Figure 11.2 Pushing position.

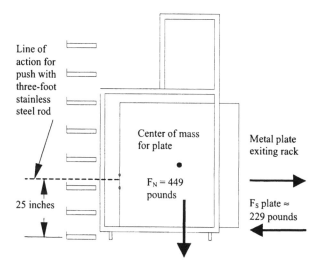

Figure 11.3. Left side view showing line of action during manual push.

11.2.1 Solution Generation Process

The injury risk level associated with this activity was high. It was apparent that other sudden injury risks such as cuts and contusions were potential problems in addition to "ergonomic injuries." For these injury-potential reasons, the issue needed to be resolved quickly. The first step in the process was to present the issue to others who might be able to generate some good ideas.

A tooling engineer and several tooling technicians reviewed workplace pictures, written documentation, and force calculations prepared by the regional ergonomist. Initial intervention ideas included: reducing friction under the plates (rollers, slick polymer, etc.); finding another well-conditioned employee to do the task; devising a mechanical assist that would accomplish the pushing (lever tool, pry bar, etc.); and modifying the top of the storage rack so that a different clamping device could be used with the existing saw shop hoist.

11.2.2 Selection of Viable Ideas

Reducing the friction would reduce the force required to push the plates out of the rack, although insufficient friction would make it difficult to control the plates. It appeared that reducing friction in the rack would be a time-consuming and expensive change. Providing an additional employee would allow the ergonomic risk exposure to be shared, however it was felt this could actually increase the injury risk potential - not a desirable option. Using a mechanical assist sounded very promising; this would focus on the specific pushing element of the technician's task and would not require the complete redesign of the saw shop. Modifying the top of the rack was not an appealing option, because more racks and rack space would be required to offset the removed rack sections. The idea of developing a mechanical assist to eliminate the manual pushing task appeared to be the most desirable option. These ideas were presented to the site ergonomics council and the saw shop technician. They agreed that pursuing the mechanical assist would be the best course of action.

11.2.3 Design Idea and Implementation

An idea was developed that incorporated a pneumatic "pusher" mounted to a rail that could be rolled into position behind a large metal plate. The saw shop technician would push the plate out of the rack by aligning the device with the plate and pushing a button (see Figures 11.4–11.5). The machine shop manager supported this idea and the decision was made to proceed with implementation. It took one week to acquire the necessary components for the device. Another full day was required for installation.

Costs related to the implementation:

1) Materials cost was approximately $40 (fasteners, a switch, air hose, angle iron, etc.). The air cylinder and industrial V track with wheels were reused from a discontinued process and were not charged to this project. It would have cost approximately $250 to purchase these items.
2) Labor was roughly $900, including the machinist's time, welder's time, and technician's time involved with clearing off and reloading the shelves. This

labor time represents a distribution of overhead time for Raytheon employees directly involved with building and installing the metal plate pusher.

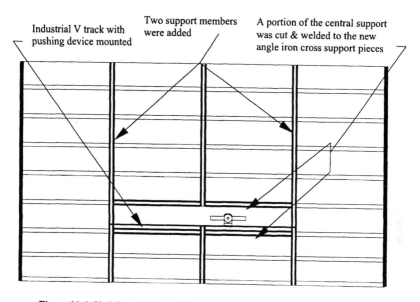

Industrial V track with pushing device mounted

Two support members were added

A portion of the central support was cut & welded to the new angle iron cross support pieces

Figure 11.4 Shelving unit viewed from the back (side adjacent to vertical rack) after changes were made.

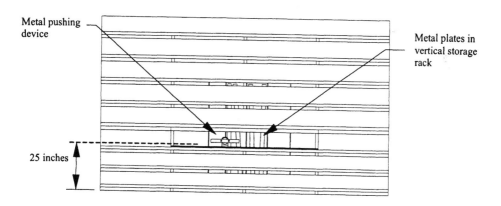

Metal pushing device

Metal plates in vertical storage rack

25 inches

Figure 11.5 Front of shelving unit after changes were made.

11.2.4 Performance of Solution

The new process for removing large aluminum plates from the storage rack merely required the technician to squat down, move the devise into place behind the plate, and depress a button (see Figure 11.6). The time to complete the pushing task had been reduced from one to two minutes down to ten seconds at most. Follow-up evaluations were performed two months and eight months after the implementation phase. Both times the equipment was performing as intended and the feedback from the technician was outstanding. The technician was very appreciative of the effort and has become a great spokesperson for ergonomic improvements in the facility.

Figure 11.6 Squatting down to activate pneumatic pusher.

11.2.5 Conclusion for Case 1

This project was clearly successful in eliminating the manual pushing of large metal plates out of the storage rack in the saw shop. The problem had been present for some time, but previous efforts had not resulted in a solution. Previous efforts had been directed at eliminating all manual-handling issues in the saw shop which were perceived to be expensive and technically challenging. Combining this with the technician's extraordinary effort to do the job successfully had resulted in a "do nothing" approach. But in the latest ergonomic analysis, the major individual ergonomic stressor was focused upon and a "low cost" solution was quickly implemented.

11.3 CASE 2 – PROBLEM DISCOVERY AND QUANTIFICATION

In 1998, a near miss involving a potential shoulder injury occurred in a precision optics manufacturing area. During an investigation of the incident, an oven loading and unloading task was identified as the primary cause. The loading and unloading tasks are performed by ten clean room technicians who must raise 15 to 35 pound lens canisters up to the spindle mounting point, which is five feet seven inches above the floor. Petite technicians, such as the five foot two inch tall woman shown in Figure 11.7, have great difficulty with the operation.

Immediate ergonomic and safety concerns included the potential for body strain injuries, and injuries related to a slip or fall while carrying the canister or leaning into the oven with the canister. Productivity concerns included: fatigue; damage to product due to forceful mating of canister with spindle ring; damage due to dropping a canister; and damage due to touching the lenses which are exposed on the bottom of the canister. Replacement costs of the lenses ranged from $5,000 to $50,000 per canister; therefore, technicians have a tremendous responsibility to handle the products carefully and will try very hard to keep product from getting damaged.

Spindle ring

Lens canister

Figure 11.7 Standing and reaching position used to load; notice standing on balls of the feet.

The technicians generated a list of factors that contributed to the stressful level of the activity.

1) The weight of the canisters.
2) Leaning into the oven, often twisting.
3) Mating the canister properly with the spindle ring. (Minor variations in the symmetry of canisters and the type of coating materials used can cause a great deal of friction between the canisters and spindle rings.)

Previous attempts to address this issue included:

1) An elevated platform to stand on. (The platform was helpful for some of the shorter technicians, but hindered those who were taller.)
2) A wall-mounted canister-lifting device next to one of the ovens. (At the end of the device was a platform that could be raised with a hand crank through a range of 18 inches. The device had two link segments and three joints allowing maneuverability in a horizontal plane. The device proved difficult to maneuver and obstructed access to the oven.)
3) A custom designed mobile lifting device quoted by an outside company for $15,000. (Management dropped the idea due to concerns about contamination of the clean room, potential damage to the products, usability, etc.)

This lifting task exceeded the basic limits stated in the lifting policy adopted by several of the major Raytheon sites in the region. That lifting policy prohibits lifting any load in excess of 28 pounds by a single individual; 18 pounds is the limit when lifting a load above shoulder height. Some exceptions exist, including when an ergonomics specialist determines that the lift is acceptable or an emergency occurs involving potential loss of life or serious injury. The lifting policy also includes training, and stresses immediate reporting of any injury. This particular canister-lifting task had been allowed due to the inability to find an acceptable alternative, although the occurrence of the near miss shoulder injury re-emphasized the importance of eliminating this manual lifting task.

A calculation using the revised NIOSH lifting equation (Water et al., 1994) was made to determine the NIOSH Recommended Weight Limit and the Lifting Index. The Recommended Weight Limit for the lift from the roll-around table to the spindle location was 24.6 pounds and the Lifting Index was 1.4. Raytheon ergonomists typically consider any lifting task in which the NIOSH Lifting Index exceeds 1.0 to be a potential problem job that may need modification or elimination.

Six factors showed that it was time for this lifting task to be modified or eliminated.

1) Exceeding the basic limits of the Raytheon lifting policy.
2) The near miss shoulder incident.
3) A NIOSH Lifting Index greater than 1.0.
4) The expressed desire by technicians to find a less strenuous loading and unloading process.
5) Ergonomist and management observations concluding that the potential for injury and product damage needed to be reduced.
6) Management commitment to pay for a "robust" alternative for this lifting task, provided that protection of both the technicians and the product could be assured through a design alternative with an effective user interface.

11.3.1 Solution Generation Process

As in Case 1, the regional ergonomist worked with tooling personnel to begin addressing this issue. Ideas that surfaced in the initial meeting included:

1) Lower the ovens.
2) Put a platform in front of the oven for the technicians to stand on.
3) Mount a lifting device to the side of each oven on which the canisters could be placed and then raised into position.
4) Design a mobile lifting device with multiple degrees of freedom to be used at all of the oven locations.
5) Have all of the canisters checked to see that they would fit easily onto the spindle rings; modify or replace canisters as needed.
6) Change the attachment method between the canister and the spindle ring.
7) Modify the spindle so that it could be mechanically lowered during the load and unload process.

11.3.2 Selection of Viable Ideas

Ideas one, two, three, and seven were quickly eliminated due to the potential expense involved or due to previous failures (ideas two and three). Ideas five and six were good ideas and should be attempted in conjunction with changing the material-handling process. The fourth idea, the mobile lifting device, appeared to be the option with the best chance of success, although it would be very challenging to make it work properly.

It was decided to create a mobile lifting device that would be able to support the canisters through the entire range of motion during the spindle loading and unloading process. The tooling team had experience building tooling for clean room environments and felt confident that an effective device could be made compatible with the particular clean room in question.

11.3.3 Expanding the Team

A meeting was scheduled with the regional ergonomist, tooling team, optics chief engineer, optics manager, site safety specialist and a representative for the clean room technicians to discuss ideas for changing the canister loading and unloading process. Everyone expressed interest in the mobile material handling device as the best option, but a number of important questions arose about how to make it work.

1) What type of lifting mechanisms would be used?
2) Would the device generate particles that could contaminate the clean room?
3) What power source would be used?
   - Pneumatic power would create concerns.
   - Certain types of electrical power sources would create concerns.
4) How much counter weight would be inside the device? Would this weight make it difficult to move?
5) How would we get around the tripping hazards associated with an air hose if pneumatic power was used?
6) How could we guarantee that the design would be able to achieve the complex articulated movement of mounting the canister with the spindle?
7) How would the device react if a jam between the canister and the spindle started to occur?
8) How well would the device protect the canisters during the movement?
9) Would there be pinch points on the device?
10) Would the device be easy to use? Would technicians abandon it for the faster, manual alternative?
11) Are we committed to make a device that will work?

The optics team was supportive of a material handling device, but they were skeptical that the device could allow for the complex interaction of the canisters and the spindle rings currently executed by the technicians, including subtle compensating actions to address irregularities in canister shape and variations in coating compounds that covered the mounting points. The chief engineer agreed to modify or replace non-circular canisters, but insisted on not changing the attachment mechanism between the canister and the ring. He wanted to see a design that could work with the existing equipment.

By the end of the meeting, the optics team was pleased that the tooling team expressed confidence in being able to create the needed device. Although skepticism remained, the optics team agreed to support of the improvement effort financially if the tooling group could present a more detailed example of a successful concept, preferably with a functioning prototype. The tooling team agreed and the development process began.

11.3.4 Designing the Prototype

Several weeks later, a pneumatically powered prototype of the canister lifter was constructed. The prototype was tested for the optics manager and chief engineer using an actual canister and tripod. The device had an articulated arm mounted to a four-wheeled cart and a spindle ring was mounted in a simulated oven-loading position. The device was maneuvered into position several times and the canister was raised approximately 25 inches and mated with the spindle. After several iterations of the maneuver, it was clear to all that the prototype functioned well.

The next step was to modify the prototype and take it to the clean room so the technicians could evaluate it under actual conditions. The actual field test of the prototype revealed several issues needing revision or fine tuning, such as better control switches, improved joints and longer link segments to allow for better canister positioning, larger castors to facilitate movement, and further adjustments to ensure clean room compliance. Suggestions came from both the technicians and the chief optics engineer.

Overall, the technicians learned how to manipulate the prototype device in just a few minutes. They liked the concept and looked forward to a fully functional model that resolved the issues discovered in the prototype test. A quote was generated by the tooling team and sent to the optics team. The quote was $4,200 for one mobile material-handling device. The majority of the cost was for labor and the $900 pneumatic cylinder that had the required 28-inch travel. Within several weeks, the optics team provided $8,500 and asked for two units to be produced.

11.3.5 Implementation of the Design

It was decided to build one lifting cart, deliver it to the clean room, let the technicians use it for a while, and then talk with them to see if any further improvements needed to be incorporated into the design of the second cart. Fabrication of the first cart took two months. Once built, a meeting was arranged with the clean room technicians for a familiarization session on how to use this new device.

Four technicians were present at the familiarization session, including the one who had experienced the near miss shoulder injury. All of the technicians were impressed with the device and appreciated the efforts to create it (see Figure 11.8).

They were asked to use it for a few weeks and keep track of any modifications that should be included in the second device.

Figure 11.8 Technician loading canister
with mobile material handling

11.3.6 Performance of the Solution

Several weeks after the initial implementation, there was a report that the device had almost tipped over while a heavy canister was being transported to an oven. Since that time, technicians were avoiding the mobile lifting device. A second meeting was held to gather technician and engineering feedback and several changes were implemented, and included reducing friction in the joints, installing a handle for pushing the device, and posting operating instructions on the device. These changes were made to the first cart and proved to be successful. The second material-handling device was then constructed.

11.3.7 Lessons Learned

As careful as we were during the initial implementation of the first device we still did not uncover all potential problems. The follow-up meeting turned out to be an excellent and necessary component of the implementation process. If it had not been performed, the project may have been a total failure with nothing to show for our efforts other than a $5,000 device sitting idle. Fortunately, the follow-up was performed and the device was a success.

11.3.8 Conclusion for Case 2

This project demonstrates a successful ergonomic improvement that reduced material handling stresses on clean room technicians during an oven loading and unloading task. The ergonomic stressor that was eliminated had been apparent for many years. Previous attempts to eliminate the problem had not succeeded due to technical challenges, combined with the perceived high cost of change and potential damage to products. A solution was found after a careful analysis of the ergonomic stressors, targeting the major stressor, providing some justification that this stressor was genuinely worth pursuing, and committing to resolve the problem.

The team approach was absolutely essential for making this project work. No single individual would have been able to solve this problem alone. Credit for the success is shared among all interested parties, because each of them contributed to the improvement. This device can also be considered a relatively "low cost" solution (as was the metal pushing device) when one considers the complicated loading task and the very high cost of the fragile lenses. Aside from the potential cost savings realized from injury avoidance, this cart would immediately pay for itself if only one canister drop using the manual technique has been avoided. Overall, this project has been deemed a success due to elimination of ergonomic and sudden injury potential along with improved task performance.

11.4 CHAPTER SUMMARY

These cases have described the successful development and implementation of ergonomic improvements in two specialized manufacturing environments. In both cases, a team-based approach was used to identify the details of the problem, generate potential solutions, evaluate, test and implement the best solution. Success arose from the cooperative effort between operators, managers, engineers, tooling personnel, site safety councils, and a regional ergonomist. The general ergonomics problems addressed had been evident for some time, but it took a structured problem-solving approach, involving all affected parties, to overcome initial skepticism and achieve workable solutions.

11.5 REFERENCES

Chaffin, D., and Andersson, G. 1991, *Occupational Biomechanics*, (John Wiley & Sons).

Flinn, R., and Trojan, P. 1990, *Engineering Materials and their Applications*, (Houghton Mifflin Co).

Waters, T., Putz-Anderson, V. and Garg, A., 1994, *Applications Manual for the Revised NIOSH Lifting Equation*, (U.S. Department of Health and Human Services).

CHAPTER TWELVE

Pork Loin Processing Lines

Tom DeRoos, Sr., CIE

12.1 INTRODUCTION

The process of preparing meat for the retail market and the dinner table requires a great amount of repetition. Conveyor chain speeds for cattle reach 400 cattle/hour, 1000 hogs/hour and 5000 chickens/hour. Large cutting tools are needed to remove bones from the carcasses. The manipulation of the tools in relation to the pieces of meat to be cut often introduces risk factors associated with cumulative trauma disorders.

Consumer demands are for further processed meats. This demand dictates that the operators remove fat, bones and prepare the meat for what is considered "case ready," or what one would purchase in stores. The design of the lines is crucial for economic production of case-ready products. This case study describes the design of a boneless pork loin line, and how the design considered the risk factors of reach, force and fatigue.

12.2 BACKGROUND

One of the most expensive cuts from a hog is the loin. This tender portion of the animal can be increased in value by having the bones removed. The average length of a loin is approximately 30 inches. Most boneless loin lines have been designed so the operator bends approximately 40 degrees forward at the waist, reaches across the product, and begins the sawing process. This reach is often excessive and places stressors on the back, shoulders, and upper extremities. Coupled with the physical stress, accurate movement of the saw requires intense mental concentration.

12.3 CHINE BONE

A portion of the backbone (chine bone) from the loin is removed using a powered saw. The saw must be manipulated in both the horizontal and vertical planes in order to remove only the chine bone. The old method required the operator to reach across a table to the distal portion of the loin, bending approximately 40°, and then pull the saw backwards while controlling the depth of the cut. Intense mental concentration and high line speeds introduce risk factors that may lead to a cumulative trauma disorder.

The new design of the saddle portion of the conveyor oriented the product closer to the operator in order to reduce both the excessive bending and the stressors placed on the upper extremities (see Figure 12.1). Aligning the loins parallel to the operator instead of perpendicular has eliminated the need to reach and bend. This has provided more recovery time and reduced the stressors on the operators.

Figure 12.1 Saw Chine Bones.

12.4 OTHER LINE CHANGES

Once the chine bone is removed, the loin is transferred to another table. To eliminate the double handling of product, the transfer incorporated mechanical means to pull the loin from the saddle onto the next conveyor. This transfer reduced elemental time constraints on the operators to orient the product and to remove the feather bones. The feather bones are removed with a straight knife and introduce little risk to the operators. The new design also allows the loin to be transferred to the new conveyor in the same parallel orientation to the operator. The risk factors reduced in this area were the reaching and bending of the operator to move the loin into place.

The button bones are actually the tips of the spine, and they need to be removed from the loin. The old method to remove these bones was to take a straight knife and make a longitudinal cut on both sides of the bones, in essence making a v-trough. This took time and also left valuable meat on the bones, reducing product yield.

A Whizard® circular knife was implemented to remove the bones quickly and leave the meat on the loin (see Figure 12.2). This increased yields and more importantly allowed more recovery time for the operators, since the Whizard® knife was faster than the manual cut.

Figure 12.2 Remove Button Bones.

12.5 CONCLUSION

Simple changes to a major pork production line created a win-win situation for this meat processor. The first and most important win was that the operators had some of their excessive motions eliminated. Some of the motions actually were high risk factors such as bending and reaching. The other win was that the product was produced in a more efficient manner and with increased yields. This is one example of how this meat processor is taking a proactive approach to applying the science of ergonomics to the workplace.

Cost savings were noted in yield savings with removing the button bones. On average the amount of yield increase has been 0.055 pounds per loin. An average plant running 8000 hogs per day realizes cost savings of approximately $110,000 per year.

Additionally, there have been no reported back or arm injuries or complaints from the chine bone saw operator since the new design was incorporated.

CHAPTER THIRTEEN

Development of a Customized Ergonomic Handle: Meat Industry Application

Dr. Greg Worrell, CPE

13.1 ABSTRACT

Hand tools assist the hands in performing a task by increasing the output force, precision, or efficiency of the hands or fingers. They can also help to protect the hands from injury, including cumulative trauma disorders (CTDs), if they are ergonomically designed. The meat hook is a tool that has become increasingly important to the meatpacking industry as the economic demands of high speed lines and the consumer demands for cuts of boneless lean meat have converged. Various changes in the design of this tool's handle, for better fit to the human hand, have been developed. The rectangular handle of the mid-1970s was gradually replaced by the cylindrical handle shape of the mid-1980s and now by the custom moldable grip handle of the mid-1990s. This latest handle design is an elliptical curve to fit the palm of the hand with a "brass knuckle" series of indentations for the fingers. The handle material not only resists slipping (important in this industry environment), but also has "recoil give" to soften the grip under heavy pulling force. Comfort of the tool has made it popular among industry employees, and the reduction of CTD symptoms has made it recommended among physicians and therapists. The design of this handle was a joint venture between the material patent holder, the tool manufacturer and an industrial ergonomist. Both the design and handle material hold promise for tools in other industries as well.

13.2 INTRODUCTION AND STATEMENT OF THE PROBLEM

The meatpacking industry has experienced a significant drop in the incidence of CTDs due to its ergonomics emphasis in the 1990s. The body segments that were most likely to develop a CTD were the fingers, hand, and wrist. The majority of meat packers in the U.S. hold a knife in one hand and a meat hook in the other. Therefore, the ergonomic fit between the handles of these tools and the hands is very important. A handle material patent holder, a hook manufacturer, and an industrial ergonomist, incorporated the handle material into the design of a handle that could be customized for ergonomic fit.

13.3 ANALYSIS

A meat hook is a desirable tool because it can eliminate a significant amount of pinch grip work with the fingers. It not only minimizes this CTD risk factor, but also decreases the chances of a knife cut by keeping the fingers further from the blade. The rectangular handle of the mid-1970s concentrated all the pulling pressure of the hook across a narrow band near the base of the fingers. The cylindrical handle of the mid-1980s spread the pulling force of the hook more evenly across the fingers, but the hard plastic handle still put contact stress on the tendons as they were squeezed against the finger bones. Therefore, circulatory compromise, neuropathy, and trigger finger CTD symptoms were more likely to develop over time. The need for a soft grip material was evident, and the ability to contour the handle to an individual's hand would be the ultimate in ergonomic design.

13.4 PROBLEM-SOLVING PROCESS

The industrial ergonomist reviewed the historical data on fingers, hand, and wrist CTDs and discussed this with physical therapists and occupational physicians in the meatpacking industry. It was agreed that the previously mentioned improvements to the meat hook would be beneficial to the reduction of CTDs to the fingers, hand, and wrist. The ergonomist then contacted Barr Brothers, a hook manufacturer, to discuss the concept and need. The manufacturer's search for an acceptable material, including USDA acceptability for food contact, led to the patent holder of PersonaGrip®. (PersonaGrip is a material that can be applied to tool handles and heated for custom forming to an individual's hand.) This team of ergonomist, hook manufacturer, and material patent holder then proceeded to work individually and collectively on the design and material blend of the new hook handle.

13.5 ALTERNATIVES AND SOLUTIONS SELECTED

The first material blend for the hook handle was too soft to be durable and did not meet USDA approval beyond the trial period. A more durable and slightly harder material blend was selected and is presently used. It still has a soft feel and give under pressure. Eight different ergonomic handle shapes (see Figure 13.1) were prototyped and distributed in the meatpacking plants for testing by the meat packers. Ergonomics or training staff at the plants collected employee feedback on the various handle shapes. Dozens of employees, performing high pull force jobs, tried the various handle shapes and ranked them according to comfort and fit. The material patent holder, tool manufacturer and ergonomist selected the current design (see Figure 13.2) from the employee input. The handle design selected is an elliptical curve to fit the palm of the hand and a "brass knuckle" series of indentations for the fingers. Additionally, the handle can be heated (with a heat gun), and then grabbed by the individual's soapy hand to custom form the handle to fit the hand. This process can be repeated if the tool is subsequently passed on to another individual.   The tool is available industry-wide through Barr Brothers Corporation of Redding, California.

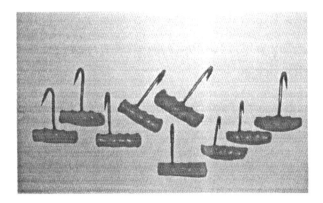

Figure 13.1 Prototype Meat Hooks.

Figure 13.2 The Barr Brothers PersonaGrip® Meat Hook.

13.6 RESULTS

The acceptance of this tool among production meat packers has been exceptional. The most common comment is that "it feels good in my hand." The handle material also resists slipping, an important consideration in the environment of this industry. Plant managers have been slower to accept the change due to the increased cost (50%), but are phasing them in. At an extra three dollars per tool, many could be purchased; this increased cost would still be less than a single doctor visit or a surgical procedure which it is designed to prevent. The occupational physicians and physical therapists are also routinely requesting the new hooks for their rehabilitation patients with trigger finger syndrome or similar CTD maladies.

13.7 SUMMARY AND CONCLUSION

This new hook with its handle's customizing ability represents the pinnacle of ergonomic design. The comfort and fit make it a popular tool among production employees. This customizing ability has applicability to the development of handles for tools in other industries as well. The significant decline in CTD incidence rates in the meatpacking industry during the late 1990s (BLS incident rate of 993 per 10,000 employees in 1998 versus 1,206 in 1995) is due to various ergonomic interventions. The ergonomic meat hook has been one of the significant changes that has helped to produce such results.

CHAPTER FOURTEEN

Elimination of the Incidence of CTDs From the Traditionally Top Risk Job In Beef Meatpacking

Dr. Greg Worrell, CPE

14.1 ABSTRACT

Meatpacking has been the top U.S. industry for the per capita incidence of cumulative trauma disorders (CTDs) from the 1980s to the present. The industry has however, experienced a significant industry-wide drop in CTDs as a result of a concerted focus in the implementation of ergonomics programs. A multitude of engineering controls have been developed to reduce the CTD risk factors of repetition, awkward posture and force. Several types of mechanical pullers, both pneumatic and hydraulic, have been developed to separate meat and bone. The job with the highest incidence of CTD in a very large beef packing plant in the early 1980s was manually pulling the "paddle bone" (scapula) from the beef carcass. Employees used their upper body musculature, as well as body weight, to pull the bone physically from the underlying meat. These repetitive and nearly maximal exertions resulted in the reporting of CTD symptoms of the fingers, hand, wrist, elbow, shoulder, and back among the majority of employees on this job. Several different mechanical pulling devices were designed over the years to address this problem successfully . As a result of these ergonomic interventions at this plant, the CTD incidence on this job in the late 1990s was reduced to zero.

14.2 INTRODUCTION AND STATEMENT OF THE PROBLEM

The Bureau of Labor Statistics (BLS) meatpacking records show 23,500 cases of repeated trauma in 1984, versus 15,900 cases in 1998. Most of this decrease has occurred in the 1990s (BLS incident rate per 10,000 employees was 1,494 in 1991 and 993 in 1998), and has corresponded with the release of OSHA's Meatpacking Guidelines. This significant drop in CTDs could be attributed to the widespread implementation of ergonomics programs and engineering interventions. When considering the labor-intensive nature of meatpacking, and that the CTD risk factors of force, repetition, awkward posture, cold and sometimes tool vibration are common, it is not surprising that meatpacking is still the leading industry for the highest rates of CTDs. Some jobs will contain all of these risk factors simultaneously, while others may have one or more risk factors at an extreme level.

Such was the case for the job of pulling the scapula bone from the beef carcass with a meat hook. The so-called "paddle bone" puller employees pulled on a meat hook with both hands. Large, whole-body exertions were often required to generate enough force to separate the bone from the underlying meat. CTD injuries were reported in the hand/wrist, elbow, shoulder and upper back. In the early 1980s this was the job with the highest incidence of CTDs for a very large meatpacking plant.

14.3 ANALYSIS

The carcasses came to the worker on an overhead chain-driven trolley where employees would use a knife and hook to separate the edges of the bone from the meat. The tip of the meat hook would then be placed in the glenoid fossa (socket of the scapula) for a maximum pull with both hands. The job analysis for "paddle bone' pulling revealed a dynamic pull force exceeding 100 pounds. A Wagner T-344137 push/pull dynamometer was fitted with a meat hook and stainless steel T-handle to make this measurement. In production, the entire job process would take about 15 seconds, and would be repeated for nearly eight hours. This would exceed 1,700 pulls per day (240 pulls per hour x 7.45 hours/day). Additional risk factors would include protective gloves on both hands for safety reasons and an environmental temperature in the 40-degree range.

14.4 PROBLEM-SOLVING PROCESS

Informal discussions were held between the ergonomist, employees, supervisors and the engineering staff concerning this job. The common consensus was that some type of mechanical assist was required to pull the bone from the meat. A primary constraint was not to damage the underlying meat in the pulling process.

14.5 ALTERNATIVES DEVELOPED

One of the simplest solutions was to loop a section of stainless steel chain around the neck of the glenoid fossa. The other end of the chain was anchored to the floor and utilized the mechanical pull force of the overhead chain-driven trolley that moved the carcass forward. As the carcass continued to move forward, the chain became taut, and the scapula was pulled free.

A second alternative developed was an air cylinder with the piston rod attached to the center of a six-inch diameter metal plate. The outer housing of the cylinder had a metal lip that was inserted under the neck of the glenoid fossa. A trigger activated the air cylinder, and the metal plate on the end of the rod pushed against the meat as it extended. This separated the "paddle bone" from the underlying meat. This mechanical puller had a handle at both the top-side and bottom-side, and was suspended by a counter-balance. The operator positioned and activated the puller.

A third alternative was a hydraulic cylinder within a metal framework on the floor. A claw-like hook (see Figure 14.1) was held by its handle and placed by the operator on the "paddle bone." A counter-balance suspended the hook. A chain connected the hook to the pulling mechanism and was activated by a foot pedal. This device is now available industry-wide through Kentmaster of Omaha, Nebraska.

Figure 14.1 The Kentmaster Hook.

14.6 RESULTS

All three alternatives were successfully used to separate the "paddle bone" mechanically from the carcass in production. The preparation of the bone with knife and hook remained unchanged, but the extremely forceful pull was eliminated. The simplest solution, looping a chain over the paddle bone, was found to work best with "hamburger grade" cattle that often had ossification at the joint. The air cylinder alternative was faster for the high-speed lines of "steak grade" cattle, but the hydraulic alternative offered more pulling power, took less skill to position, and was less likely to slide off the bone. The hydraulic alternative was therefore the method selected. Several claw-like hook configurations have been developed over the years.

The CTD incidence on this formerly high turnover job dropped 100%. It was once the highest CTD job in the plant, and was reduced to zero cases. Current cost of the mechanical puller ($2,500 air or $5,000 hydraulic) is less than the medical and worker's compensation costs saved in a year. Staffing per line-speed was not changed. The job is now simple enough, and lacking in CTD risk factors, that it is often used as a restricted-duty job assignment.

14.7 SUMMARY AND CONCLUSION

The total elimination of the incidence of CTDs from what was formerly the top CTD job in a large beef meatpacking plant was accomplished by the development of mechanical assists. These mechanical "paddle bone" pullers greatly reduced the near maximal manual pulling associated with the separation of the underlying meat from the scapula. The development and implementation costs of these ergonomic assists were more than offset by the savings in medical and worker's compensation costs.

CHAPTER FIFTEEN

Ergonomic Redesign of a Hanging Table for Poultry Processing

David J. Ramcharan, Ph.D., CHFP

15.1 INTRODUCTION

According to the Bureau of Labor Statistics in 1980, repetitive trauma disorders (RTD) represented 18% of OSHA-recorded work-related illnesses. This number grew dramatically over the following 10 years, and by 1990 it was up to 55%. Although many thought that this growth was spurred by changes in record-keeping requirements and the changing definitions of a work-related RTD, OSHA could not ignore the statistics.

In 1991, the Occupational Safety and Health Administration (OSHA) first published the "Ergonomics Program Management Guidelines for Meatpacking Plants," commonly called "The Red Meat Guidelines." These guidelines were focused on the meatpacking industry because of that industry's high incidence rates compared to the national average. Its major sections were management commitment, worksite analysis, hazard prevention and control, and medical management and training.

Since the Red Meat Guidelines were specifically targeted at the meat processing industry, most major poultry processors developed and implemented ergonomic management programs long before other industries started considering ergonomics.

15.1.1 Poultry Processing Industry

The poultry processing industry is similar to many assembly line manufacturing industries in that employees are generally assigned to a specific position along a line and are given a specific task to do.

There are, however, many characteristic differences that should be noted.
- Jobs generally include either cutting or pulling meat.
- While many of the jobs are automated, the loading and sometimes unloading of the machines are generally manual.
- Because this is a food item, all employees are required to wear gloves.
- By its very nature, fresh meat is slippery.
- Since meat is a perishable item, the USDA requires the meat-handling areas to be kept below 40°F.

Many of the above characteristics are directly related to the "risk factors" for the development of musculoskeltal disorders (MSDs) listed in OSHA's guidelines.

This paper deals with Hazard Prevention and Control at one specific poultry operation, the hanging table. The focus is on the principles and process of analysis and design that were used.

15.2 DATA COLLECTION

In order to control the incidences of MSDs in the workplace, accurate and timely data collection and analysis must be performed.

Since, at the time, there was not a computer network installed at this facility that would allow the computerization of this process, a paper flow had to be defined to support this data collection.

First, it had to be decided how the data were going to be used; that is, what questions is it supposed to answer. For our program, we decided that it was not good enough to track incidences of MSDs. A management commitment was made to try to identify any employees with MSD-type symptoms before they were officially defined as an MSD case statistic.

Secondly, since we had two shifts of people doing the same tasks, we also wanted to track any groups of employees doing the same jobs who experienced the same symptoms, even if they were on different shifts.

To accomplish these two goals, the medical and safety staff designed treatment logs that would be kept by the on-site nurses. In order to minimize the nurse's workload, these forms also included spaces to allow the nurses to enter all required medical and OSHA information for every on-site nurse station visit. A new form was used for each new shift. Since there were many processing plants, the location of each facility was also noted on the form.

Of all the data included on the form, the relevant information needed for the analysis were:
-    department
-    job position
-    body part affected
-    symptoms reported
-    treatment.

Table 15.1 shows an example of these relevant columns from this log. For the sake of medical privacy, the employees' names were not released to anyone outside of the medical staff.

These visits were tallied every day by department. Analyzing the data and identifying trends then became a straight-forward task of following the number of visits per shift by department. Using this approach, day and night shifts could also be compared to identify any significant differences that might exist. Table 15.2 gives an example of the tally form used for data collection. Depending on the incident history, additional columns can be added for symptoms related to hips and ankles.

Table 15.1 Necessary columns from nurses' logs.

Location: Shift: Date:				
Department	Job Position	Body Part	Symptoms	Treatment

Table 15.2 Example tally form.

Location: Shift: Date:							
Job Position	Neck	Shoulder	Back	Elbow	Wrist	Finger	Knee

15.3 PROBLEM IDENTIFICATION

The concern with the hanging-table was identified by the nurse on the night shift. It was noticed that there were daily visits of all employees that worked at that table all doing basically the same job. Every day the employees would come in before the shift and wrap their elbows with ace bandages. When this continued for more than a couple of days, the nurse started to probe into the reasons for the wraps.

As it turned out, all of the employees doing the hanging job at that workstation were experiencing some mild discomfort in their wrists and elbows. After one of those employees found out that the wraps helped to relieve the discomfort, the wrapping spread to the rest of the group.

Upon identifying this, the nurse then requested that the ergonomist analyze this job to see if anything could be done to help relieve these symptoms.

Analysis Concept

If multiple employees from varying backgrounds have the same symptoms when doing the same task using the same methods and equipment, then the design of the methods and equipment might be contributing to those symptoms.

15.4 ANALYSIS AND DESIGN

The steps used to perform the analysis of the hanging table and the resulting design included the following.
- Data collection:
  - Direct observation of the work methods.
  - Measurement of the original workstation.
- Analysis:
  - Identification of the workstation characteristics that are of concern.
  - Identification the anthropometric measures needed for the design.
- Design:
  - Selection of the anthropometric measurements.
  - Development of an initial design.
  - Prototyping (verifying) the design.

15.4.1 Data Collection

The following work method was observed from direct observation of the employees. Chickens are presented on a conveyor passing in front of the employee. (The "table" in hanging table is actually this conveyor.) Three employees stand on one side of the conveyor (chickens flow left to right) and three stand on the other side (chickens flow right to left). Figure 15.1 shows the original dimensions of the workstation. For the purposes of this description, we will assume the chickens are flowing left to right.

1. A chicken is picked up with either hand with the palm facing downward below waist level.
2. The chicken is then lifted to about chest level and the hand turned over (now palm up.)
3. The chicken is then hung on the shackle by reaching forward and catching the knuckle of the leg in the shackle.

15.4.2 Analysis

As in most material handling tasks, the key workstation characteristics of concern were the pickup location and the release location. For this particular task, the pickup location was the conveyor top, and the release location was the shackle. The measurements for these were easily identified from the original workstation's dimensions.

Before the anthropometric dimensions could be identified, the postures of the joints involved in the task needed to be specified. Since this hanging task was done with the worker standing, the ergonomic design guidelines that applied were as follows.

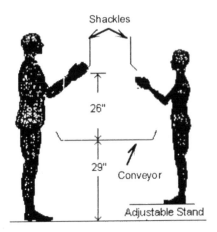

Figure 15.1 Original workstation.

Ergonomic Design Guidelines

1. Design for a straight-forward facing posture.
2. Design for an upright posture.
3. Keep the activity of the joints around the midpoints of their range of motion.
4. Keep the hands below shoulder level.

For this task, these guidelines translate to:
- Keep the employee from bending to pick up the chickens; that is, the minimum pickup height should be no lower than the knuckle height with arms extended downward.
- Keep the employee from having to raise the hands above shoulder level; that is, the maximum height of hands should be no higher than shoulder height in an upright posture.

Since our goal was to accommodate at least 90% of the population (using data from U.S.), both the 5th percentile and 95th percentile measurements were considered. Figure 15.2 shows the applicable postures with the corresponding measurements (Eastman Kodak Company, 1983).

15.4.3 Design

Given these anthropometric measurements and the task description, the next step was to select which measurements should be used in the design.

Ergonomic Design Guideline

   5.   Allow the shorter person to reach and the taller person to fit.

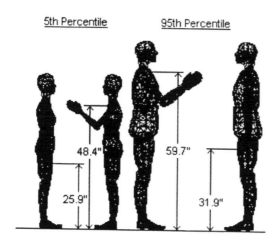

Figure 15.2 Applicable anthropometrics.

Ergonomic Design Guideline 5 was applied by first identifying which measurements could be considered "reaches" or "fits."

As mentioned earlier, the standing knuckle and shoulder heights were the main dimensions that were considered. Since shorter employees can use adjustable stands, the shackle height selected was the shoulder height of the 95th percentile person. This provided a "fit" for the taller worker, and it was reachable by the shorter worker standing on the adjustable stand. Now, assuming shorter employees had their shoulders at shackle height, if the distance between the conveyor and the shackle was designed for the 95th percentile measurement, the distance from pickup to release would be too large for most of the population. This was considered a "reach," and the 5th percentile measurement was used for this dimension to create a "fit" condition.

Design Definitions

A "reach" occurs when a dimension, if made larger, does not allow a person to maintain a neutral posture while trying to touch an object at that dimension.

A "fit" occurs when a dimension accommodates for a person's size without requiring non-neutral postures.

The resulting design set the bottom of shackle height at the 95th percentile shoulder height (approximately 60 inches). The design set the distance between the bottom of the shackle and the conveyor top at the 5th percentile measurement (approximately 22 inches). See Table 15.3.

Table 15.3 Coarse Measurements for Design.

	5th Percentile (inches)	95th Percentile (inches)	Selected Measurement (inches)
Release Height (Bottom of Shackle)	48.4	59.7	60
Max. Conveyor to Shackle Distance	48.4 - 25.9 = 22.5	59.7 - 31.9 = 27.8	22

Given these dimensions, the next step was to determine how to adapt the 5th percentile person to the 95th percentile bottom-of-shackle height.

Adjustable Design Principle

For height variations, design for the largest and give shorter people adjustable stands.

Since the difference in shoulder height between the 5th and 95th percentile measurements is approximately 12," the approach taken was to provide a stand that was adjustable in 1" increments, up to a maximum of 12" above ground level. This adjustability allowed for all workers who were within the 5th and 95th percentile measurements to reach the shackles.

Since it is costly to implement changes on a live production line, this initial design had to be verified before it was implemented. This involved revisiting the site and making sure all of the dynamics of the task and product flow could be accommodated by the design. The main issue to be addressed was the clearance needed between the chickens hanging on the shackles and at least one layer of chickens on the table, or conveyor.

To address this issue, the range of the sizes of chickens being handled needed to be determined. This was done by collecting measurements from random samples of chickens. The data showed that the measurements for all of the birds had very little variance. This made the design adjustments much easier than expected.

The width (3") and the length (13") of the chicken were important because they both contributed to the clearance needed for one layer of chickens on the table and one hanging from shackles passing above it.

Since the designed clearance was 22," this allowed for the 16" space needed, and still allowed the workers to reach under a hanging chicken to pick up another chicken without bumping the passing chicken. Additionally, the width of the body of the chickens (3") was important because the chicken was generally picked up by the thigh that projects above the body. This produced a beneficial side effect. It allowed the actual height of the hands to be 3" higher at pick-up than initially designed. This made the pick-up closer to the more ideal neutral elbow-height position of the hands. Figure 15.3 shows the dimensions of the new workstation design.

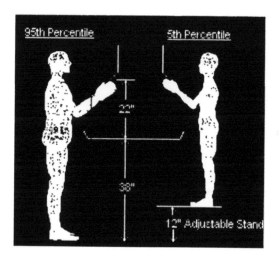

Figure 15.3 Dimensions of the new workstation design.

15.5 SUMMARY

Fortunately, the facility in which this hanging table was located had its own fabrication department. Also, there was a great deal of management support for the ergonomics program. Therefore, the recommended workstation dimensions were quickly implemented.

Immediately after the design was implemented, the symptoms that the employees were experiencing in their elbows subsided. Even after 3 months of follow-up, there were no MSD concerns or symptoms experienced at this work station.

As an ergonomist, I was certainly pleased with the success of this redesign. Even though the ergonomics program had management support, this success gave the program a much needed credibility boost with production managers.

During my observations, I noted that the employees were, in some cases, throwing the chickens at the shackles. This resulted in approximately 50% of the chickens falling back on to the table. This technique was probably developed as a way to compensate for reaching to the shackle that was placed too high. With the redesign, one benefit was that repetitions decreased by 50%, since the employees could easily reach the shackles and changed their habit of throwing the chickens at the shackle. I am sure this also had a significant role in reducing the symptoms.

Although this paper highlights Worksite Analysis and Hazard Prevention and Control efforts, in most cases, the full benefits of MSD management cannot be achieved without appropriate management commitment, medical management and employee training.

Although meat processing is certainly different than other industries, I am certain than much can be learned by studying the meat processors' experience with MSD management. Because of the mixed blessing of early attention by OSHA, I believe the poultry industry is one that has led the way for other industries in the industrial ergonomics management approach to the control of MSDs.

15.6 REFERENCES

Eastman Kodak Company, 1983. *Ergonomic Design for People at Work*, Volume 1, Appendix A.

R. Blake Smith, January 1993. OSHA Gathers Information for First Draft of National Ergonomics Safety Standard, In *Occupational Safety & Health*, January 1993, pp. 46-49.

SUPPORT
AND
SERVICE JOBS

CHAPTER SIXTEEN

Bottled Water Handling: A Case Study Through Biomechanical Analysis

John P. Cotnam, CSP, Chien-Chi Chang, Ph.D. and
Victor S. Garrison, CPE, CSP

Abstract

There are numerous physical hazards present in bottled water delivery operations that may contribute to acute and chronic soft tissue injuries. This chapter summarizes ergonomic analyses of joint loading considerations (e.g., forces and moments) while handling bottled water during delivery operations. Field survey videos of bottled water handling tasks were collected. The joint loading, joint strength limits, and L5/S1 compressive force for each individual task were estimated by using the VidLiTeC™ software analysis system. A simplified biomechanical model for lateral bending moments estimated at the hip joint while carrying the bottled water was also used. The results of this biomechanical analysis show how design changes to minimize overhead reaching and bending while lowering bottles from the delivery truck can reduce joint loading. Carrying one bottle on the shoulder versus hand carrying can also reduce lateral bending moment at each hip.

Introduction

Handling bottled water (lifting, lowering, pushing, pulling and/or carrying) represents significant physical exposures for work-related musculoskeletal disorders. The process of transferring water bottles from the delivery truck to the customers' premises can be described in three basic steps, 1) physically lowering the required number of bottle(s) down from the delivery truck and placing them on the ground or on a hand cart, 2) manually carrying the bottle(s) or alternatively, using a material handling device such as a two-wheel hand truck or a four-wheel cart to push/pull the bottles into a delivery site, and 3) lowering or lifting the bottles at the customers designated storage area inside (e.g., office storage closet). Lowering bottles down from the delivery truck requires the driver to reach overhead with one or both arms to grasp the bottle by the neck, pull it out from a crate or rack, and then lower the bottle to the ground or a material handling aide.

The typical design of the delivery truck requires the driver to lower bottles using awkward postures. The trucks used for the delivery of spring water have numerous bays on either side. Within these bays, bottles of water are stacked vertically in crates or racks. The height of the top bottles may exceed 80" on many truck designs which requires even the tallest drivers to extend to their maximum reach limit in order to grasp the necks of the bottles.

Delivery drivers will commonly carry one or two bottles of spring water from the truck into the delivery site. For a one bottle delivery, the worker might carry the bottle straight down grasping the bottle by the neck, or balanced on his or her shoulder. For a two bottle delivery, the worker will either carry both bottles straight down, or one bottle straight down and the other balanced on his/her shoulder. This biomechanical assessment focused on joint loading which occurs while lowering bottled spring water from a delivery truck and carrying bottled water using different methods.

In general, four different approaches have been employed to study manual material handling tasks: the biomechanical approach, the epidemiological approach, the physiological approach, and the psychophysical approach. The goal of this evaluation was to estimate joint loading considerations and compare these to the strength capacity of joints or tissue tolerances where possible. Therefore, the biomechanical approach was used.

2.0 Methods

Two different biomechanical analyses were performed. The first analysis addressed the para-sagittal joint loading which occurs while lowering bottled water from a truck. A biomechanical analysis model (Chaffin 1991) often used by ergonomists was applied. The VidLiTeC™ software analysis system was used to estimate the joint loading, joint strength limits, and L5/S1 compressive forces during different tasks. The second analysis addressed the lateral joint moment applied on the hip joints of an individual carrying bottled water using different methods. A simplified biomechanical model was developed to estimate the lateral bending moment of the hip joints. The effects of different bottled water carrying methods are discussed.

2.1 Joint loading analysis in sagittal plane

When performing a biomechanical analysis, the information of the body segmental coordination is crucial. Specialized motion tracking equipment must be used to collect kinematic data during lifting and lowering tasks. However, due to the complex setup of both hardware and software, such equipment is usually suitable in the laboratory environment only, and has limited appeal for a field application.

A video-based lifting technique coding system (VidLiTeC™) (Hsiang et al., 1998, Chang et al., 1999) was proposed and developed to help ergonomists conduct biomechanical analyses in the field environment. By using videotapes from a field survey and inputs of the subjects' gender, weight, height, and weight of the load, the system can assist the ergonomist in performing such analyses without any special equipment to collect or track subjects' joint trajectory information. The major advantage of VidLiTeC™ is that it allows trained users with simple equipment (camcorder) to collect large sets of lifting data while maintaining the reasonable quality of analysis results.

The details of the VidLiTeC™ system are not discussed in detail. Those readers interested in additional information are encouraged to refer to the given references.

Four subjects (actual drivers) participated in a field simulation of lowering bottled water from a delivery truck. Subjects were videotaped lowering bottles from three levels (second, third and fourth highest bottles) inside a full bay of the delivery truck. The vertical heights of the bottle necks (first grasping point) from the ground were 118, 150, and 182 cm. The four subjects lowered the bottles using a two-hand technique. The VidLiTeC™ system was then used to estimate the joint moments during these tasks.

The motion video was captured and digitized into the computer program individually, for each task. A simple interactive dialog box requiring the user to input the subject's basic information is shown on the left side of the screen shot in figure 1. The right side of this screen has a window that displays the captured video clip.

After the four events are selected, the user manipulates the 3D mannequin in the interactive window to help identify the major joint angles (figure 3). Through the 3D computer graphical interface, all four postures from the four essential events were adjusted repeatedly until the result was satisfactory. Upon completion of the initial steps, the computer program then generates the motion pattern according to the input parameters. The joint motion patterns, moments, strength limits, and L5/S1 compressive forces are estimated based on the embedded biomechanical model (Chaffin et al., Winter et al.). Estimated individual anthropometric information (e.g., segment weight, length, center of mass, and moments of inertia) and motion pattern prediction figures are generated as part of the system output capability.

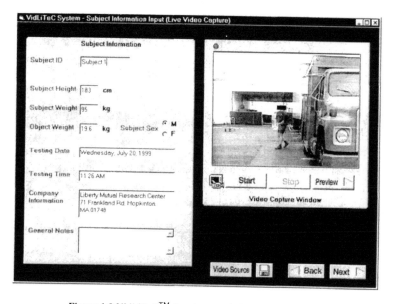

Figure 1.0 VidLiTeC™ data input and video capture screen

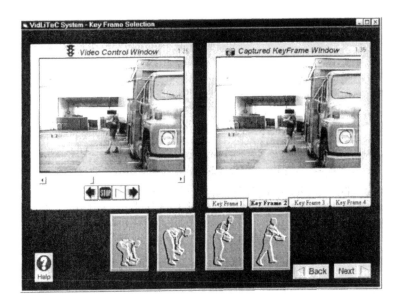

Figure 2.0 VidLiTeC™ key frame selection

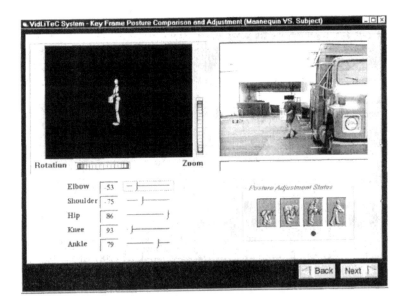

Figure 3.0 VidLiTeC™ Postural adjustment of the 3D Mannequin

2.2 Lateral bending moment at the hip

In addition to the biomechanical analysis in the para-sagittal plane, the effects of carrying the bottled water on the lateral bending moments at each hip joint was evaluated. A simplified static biomechanical model used to estimate the bending moment in frontal plane is shown in figure 4.

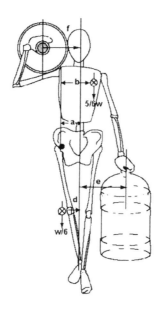

Figure 4.0 Lateral bending moment model

Figure 5.0 shows the different carrying methods assessed using this approach. The formulas of lateral bending moment (Nm) at the hip joint in response to each different carrying method are expressed in table 1.0, where W is the body weight (784 N), B_w is the weight of bottled water (196N), and the parameters a, b, d, e, f are the corresponding segmental lengths and distances from the center of mass of each link to the center line as shown in the figure 4.0.

The dynamic effects were not considered in this simplified model. The model assumes that the leg opposite to the lateral side being evaluated is raised slightly to simulate conditions during normal gait.

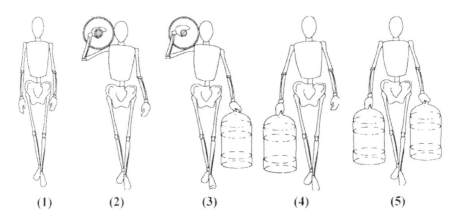

(1)   (2)   (3)   (4)   (5)

Figure 5.0 Bottled water loading conditions (1= no load, 2= shoulder, 3= shoulder/hand, 4= hand, 5= hand/hand)

Table 1.0 Lateral bending moment equations

Condition	Right Hip Moment	Left Hip Moment
1	$M=5/6*W*b$	$M=5/6*W*b$
2	$M=(5/6*W*b + (a-f)*B_w)$	$M=(5/6*W*b + (a+f)* B_w)$
3	$M=(5/6*W*b + (a-f)* B_w + (a+e)*B_w)$	$M=(5/6*W*b + (a+f)* B_w + (a-e)* B_w)$
4	$M=(5/6*W*b + (a-e)* B_w)$	$M=(5/6*W*b + (a+e)* B_w)$
5	$M=(5/6*W*b + 2*a* B_w)$	$M=(5/6*W*b + 2*a* B_w)$

Substituting each subject's anthropometric data (weight and segment lengths), distance from the bottle center of mass to the subject's center line, and the weight of the bottled water into the formulas shown in table 1, we can estimate the joint bending moment based on the different handling conditions.

3.0 Results and Discussion

3.1 Joint loading analysis in para-sagittal plane

Subjects were videotaped lowering bottles from the 150 cm rack slot height to a hand cart (50cm off the ground). The average peak L5/S1 compressive force while lowering bottles onto a hand cart was 3,509 N as compared to the average of 4,002 N lowering bottles to the ground. Lowering bottles to the ground versus a hand cart increases the peak average compressive force by over 14%. Therefore, drivers should be instructed to use the hand cart whenever practical during delivery operations.

In general, drivers tend to carry bottles to avoid the extra time required to deploy and store the material handling aide even on deliveries where using the device is practical. Therefore, management is challenged to influence the decision making of drivers so that material handling equipment are used when appropriate. The added time it takes to use this equipment should not always be viewed as a cost.

Figure 6 shows the predicted joint moments (dark line) at the elbow, shoulder, hip, knee and ankle. The lighter line represents the joint strength limit (Stobbe, 1982). Examining the figure for the shoulder, we see the highest predicted moment occurs in the first quarter of the lowering sequence (e.g., when the bottle is lowered from the truck to the position closest to the subject's body). Truck redesign of the rear bay to eliminate the need for excessive overhead reaching can reduce the joint moments at the shoulder during the first phase of the task. Instructing drivers to stand close to the bottle on the truck before it is lowered is also expected to reduce this moment during the first phase of the lowering task.

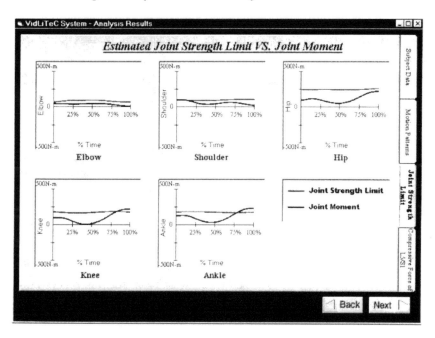

Figure 6.0 Joint moments and joint strength limits at elbow, shoulder, hip, knee and ankle.

VidLiTec™ assumes the manual lifting/lowering task is performed with two hands, and all subjects in this study chose this technique. However, some drivers in industry have used a one hand method to lower bottles from the truck. It is expected that one hand lowering methods would create excessive moments at the elbow and shoulder based on the results from this investigation. Therefore, training efforts to instruct drivers to always use the two hand method should be emphasized.

Maximum knee and ankle joint moments (refer to figure 6) occur during the final quarter of the lowering task. Keeping the bottle close to the body while it is being placed onto a handcart or the ground is expected to reduce these moments (possibly below the joint strength limits) as well as improve the subject's balance during this final phase of the lowering task.

3.2 Lateral bending moment at the hip joint

The estimated joint moment at each hip across the four different bottled water handling methods was performed using the equations shown in table 1. The lateral moments (Nm) across the five loading conditions studied are shown in figure 7 below for one subject.

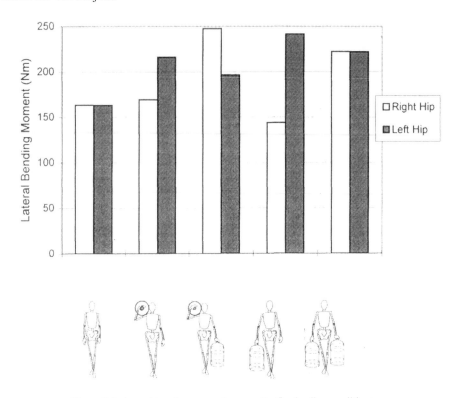

Figure 7.0. Lateral bending moments across the five loading conditions

Referring to the calculated lateral bending moment at the hip joints in table 2.0 we see that the greatest difference in moments between hip joints occurs in condition four (one bottle hand carried). Carrying one bottle on the shoulder results in a smaller peak moment in comparison to hand carrying. Therefore, this method is preferred when considering just lateral bending moments.

Table 2.0 Bending moments and loading at the hip across five conditions

Condition	Joint Loading* (Kg.)	Bending Moment (Nm)		Δ Bending Moment (Nm)
		Right	Left	
1	0	163.3	163.3	0
2	19.7	216.3	169.2	47.1
3	39.4	196.7	247. 6	50.9
4	19.7	241.7	143.7	98
5	39.4	222.1	222.1	0

* Loading due to bottle weight only

Maximum joint loading due to the external force of the bottle(s) occurs when two bottles are carried (conditions three and five). Comparing these conditions, hand carrying two bottles results in equal moments at both hips (e.g. Δ bending moment=0) and a smaller peak moment. Therefore, two hand carrying is preferred to hand and shoulder carrying based on just this biomechanical criteria.

Other considerations relative to gait stability under the various loading conditions should also be considered when determining the best technique for a given delivery situation. From a physiological perspective (outside the scope of this case study), two bottle carrying techniques may be more fatiguing than one bottle handling techniques. A detailed physiological analysis of the effects of carrying water (e.g., number of bottles required to be handled, distance the bottles must be transferred, grade of the walk, etc.) under these four loading conditions would be necessary.

This case study illustrates the importance of using the biomechanical approach to quantify joint loading considerations for different lowering and carrying tasks related to the handling of bottled water. Results can be used to in conjunction with other criteria to justify equipment redesign as well as define safe and efficient delivery methods which reduce musculoskeletal disorders.

References

Chang, C-C, Hsiang, S.M., McGorry, R.W. 1999, Segmental Coordination in Lifting Tasks, Advances in Occupational Ergonomics and Safety, 85-88.

Hsiang, S.M., Brogmus, G.E., Martin, S.E., and Bezverkhny, I.B. 1998, Video based lifting technique coding system, Ergonomics, Vol. 41, No.3, 239-256.

Chaffin, D. B. and Anderson, Gunnar B. J. 1991. Occupational Biomechanics, John Wiley & Sons, New York.

Stobbe, T. 1982. The development of a practical strength testing program in industry, unpublished Ph.D. dissertation, U. of Michigan, Ann Arbor.

Winter, D.A. 1979, Biomechanics of Human Movement, John Wiley & Sons, New York.

CHAPTER SEVENTEEN

Work-Related Musculoskeletal Disorders: Ergonomic Risk to Healthcare Workers

Alison G. Vredenburgh, Ph.D., CPE and Ilene B. Zackowitz, Ph.D.
Vredenburgh & Associates, Inc.

In the cafeteria of the Green Forest Medical Center*, a group of employees takes a break. Nurse Ivy commented to hospital tech., Suzie Aiken, "My neck is so stiff, I don't think I can finish my 14 hour shift." "Oh Yeah?" said Suzy, "I just helped haul up a 500 lb. patient off the floor. She was too big for any of our lifts. I don't think I'll be able to walk upright anytime soon." Marge Broomswift, one of the hospital's housekeepers shakes her head saying, "I've got to move those beds with around without help. Most of the wheels on the beds are broken. At least nurses move beds together. They expect housekeepers to move them alone. Besides, I've changed so many beds today, I've lost count. Now, I can't even feel, let alone use, my wrists."
*Fictitious name

Why are healthcare workers particularly susceptible to musculoskeletal injuries?

What are the most prevalent types of injuries sustained by healthcare workers?

Which types of healthcare tasks are most likely to produce work-related musculoskeletal disorders (WMSDs)?

What specific steps can hospitals take to reduce the rate of injuries sustained by healthcare workers?

1.1 WHY ARE HEALTHCARE WORKERS PARTICULARLY SUSCEPTIBLE TO MUSCULOSKELETAL INJURIES?

According to the National Safety Council (1997), there were 3.9 million work-related injuries in the U.S. in 1996 alone, costing taxpayers a staggering $121 billion. Sadly, an estimated 17 American workers die each day from injuries sustained while on-the-job. These employee injury rates continue to rise and show no signs of abating.

Unfortunately, healthcare facilities are not immune from either the high rates of work-related injuries or the associated economic and staffing consequences that follow. U.S. Department of Labor statistics indicate that the rate of job-related injuries sustained in hospital settings is 6.2 cases per 100 full-time workers annually. Of these, musculoskeletal disorders are the most prevalent and costly type of injury. The average claim costs $24,000, and the workers compensation costs for healthcare employees exceed $1.5 billion annually.

Currently, there are nearly five million employees working full and part-time in hospitals and related health service occupations (e.g., nursing homes and home-health). Many of these people work under difficult and stressful circumstances. Factors that contribute to unsafe conditions in healthcare environments include unpredictable shift rotations and the strain caused by a frequent requirement to work overtime. In light of the poor working conditions present in healthcare environments, it should come as no surprise that 15 of the 27 occupations ranked highest in stress are in healthcare. Since hospitals never shut down, and indeed, rarely slow down, this becomes a setting condition for the occurrence of employee injuries that is exacerbated by a number of factors, including fatigue, stress, and policies that contribute to unsafe working conditions.

Although it might seem reasonable to expect that occupational safety and health services would be conveniently accessible to healthcare workers, this is rarely the case. In a study conducted by the National Institute for Occupational Safety and Health, only 8% of the 3,686 hospitals surveyed met all of NIOSH's basic components of an effective occupational safety and health program for hospital employees. One possible reason for the lack of services available to healthcare workers is that hospital administrators may not consider employee health to be a high priority. Even if they do, however, they may view the need for a special health program for hospital employees as redundant, since healthcare is the main business of hospitals.

Improving the quality of healthcare that is available to healthcare workers might best be achieved by focusing on improving aspects of the healthcare delivery process. Existing evidence clearly shows that hospitals that incorporate quality improvement methods into their procedures achieve higher levels of efficiency, patient satisfaction, safety, clinical effectiveness, and profitability than those that do not. However, because of the traditional focus on sick care, rather than on prevention, and because of the high initial costs of developing effective safety and health programs, this approach to improving workplace safety is rarely adopted. Instead, administrators tend to opt for implementing superficial employee training programs that cover only required basics, such as communication and problem-solving skills and the use of quality measurement techniques. Unfortunately, quality measurement tools tend to unveil the flaws, and not their causes.

1.2 WHAT ARE THE MOST PREVALENT TYPES OF INJURIES SUSTAINED BY HEALTHCARE WORKERS?

The types of safety hazards present in hospitals are well known and include strains and sprains, needle punctures, communicable diseases, toxic and hazardous substances, dermatitis (caused by handling cleansers, medicines, antiseptics and solvents), and thermal burns (primarily in food service, laundry and sterilizing areas). According to Cal/OSHA (1997), back sprains and strains are much more common among hospital workers than among people in other types of occupations. Their data show, for example, that 46% of nurses, aides, orderlies and attendants report suffering back injuries, compared to 26% in private industry occupations.

Slip and fall accidents are prevalent in all types of work environments and cost an average of $28,000 for each lost-time injury (injuries severe enough to cause an employee to take time off from work). The National Safety Council indicates that same-level falls (no change in elevation such as down stairs) account for nearly 16% of all accidents that result in lost workdays.

Vredenburgh (1998) recently reviewed employee injury data provided by 62 hospitals over a 3-year period. The types of injuries identified in that study are summarized in Table 1.1. As evident in this table, work-related musculoskeletal disorder injuries were the most prevalent type of injury sustained.

Table 1.1. Mean Frequency of Each Injury Type, Percentage of Total Frequency and Mean Frequency Per Year Per 100 Employees.

Injury type	Mean frequency/ year	Percent of total	Mean frequency per year per 100 employees
Work-related musculoskeletal disorders	47.8	40	7.4
Needle punctures, blood exposure	14.9	13	2.7
Contusions	14.9	13	2.1
Lacerations/cuts	10.4	9	1.9
Other (allergic reactions, unknown causes)	6.5	5	0.9
Disease exposure	5.4	5	0.7
Burns	3.2	3	0.6
Abrasions	3.8	3	0.6
Eye injuries	3.1	3	0.6
Skin disease	2.9	2	0.4
Finger injuries	2.2	2	0.3
Toxic exposure	2.0	1	0.3
Mental stress	0.9	1	0.2
Human or animal bites	0.5	0	0.1
Total	**118.11**	**100**	**18.8**

1.3 WHICH TYPES OF HEALTHCARE TASKS ARE MOST LIKELY TO PRODUCE MUSCULOSKELETAL DISORDERS (WMSDs)?

1.3.1 General Hospital Environmental Factors

Hospital healthcare workers are commonly pulled off one job to perform another task, which in turn, results in a significant backlog of work that must be completed.

When time pressure becomes a central issue, and numerous responsibilities must be addressed simultaneously, healthcare workers may be forced to "cut corners" in order to get the job done. Predictably, this combination of factors can lead to mistakes, and in turn, to a significant proportion of on-the-job injuries in these settings. It is noteworthy that many employees claim that until fairly recently, rushing was less of a contributing factor to accidents because more sufficient levels of staffing were present.

1.3.2 Lifting and Moving Loads

1.3.2.1 Lifting and moving patients

The majority of workers compensation costs are the result of injuries caused by moving or transferring patients. This may be due, at least in part, to the fact there are no guidelines regarding how many people are needed to lift a patient of a certain size or weight. However, a number of other factors also seem to play a role. One factor is the sheer bulk of the patients. Currently, the average weight of hospital patients is about 200 pounds, which is significantly more than the average person can safely lift. Yet, healthcare workers frequently attempt to do so, since it can be difficult to assemble—on demand—the people needed to safely lift or transfer a heavy patient. Moreover, not all hospital employees are willing to help; doctors are sometimes willing to assist in moving patients, whereas housekeeping personnel are usually not (or they are not asked). A related factor is that employees' job descriptions and duties remain the same (including lifting), regardless of their age.

Lifting injuries frequently occur when patients who are not supposed to be out of bed attempt to get out by themselves and begin to fall. When this happens, hospital employees are injured while attempting to help the patient recover from the fall. Patients increase the potential for injury when they become combative and fight the help they are receiving.

The lack of (working) equipment or poorly designed equipment also contributes to lifting-related worker injuries. For example, many beds do not stay put when moving patients because the brakes do not work. And some beds are not compatible with the overhead equipment designed to help patients support their weight during transfers. Finally, there are architectural problems to consider. Showers, for example, often have high lips that are difficult to maneuver patients over when trying to bathe them.

1.3.2.2 Moving of furniture, beds, and other equipment

When moving furniture, sometimes a dolly cannot be used due to the building layout (such as stairs with low ceilings); in these cases manpower must be used. At times, employees are instructed to move full filing cabinets. Beds often have parts (part of the frame/mattress) that have to be supported while cleaning the bottom of the bed. Many beds are heavy and their brakes often do not properly release; thus, requiring a lot of effort to move. Beds often have a wheel that is not aligned or does not roll. At times one employee must push a bed alone. Employees frequently have to lift and carry large heavy supplies (such as bleach containers or boxes of gloves) from high shelves.

Photograph 1.1 Beds often have heavy parts that must be supported while cleaning under the mattresses.

1.3.2.3 Trash and laundry management

Laundry and trash bags are often overfilled such that they are too heavy to carry safely; employees may decide to drag the heavy bags down the hall. Often the bags are hard to remove from the bins because the bags have suction with the bin. There is difficulty pulling wet linens out of washers; sheets and towels often get tangled in the large capacity washers. For hospitals that have their own laundry facility, laundry workers are exposed to many WMSD risk factors due to the volume of linens and the large size of the machines. There is difficulty throwing wet items to bins and from bins to dryers, as well as lifting the heavy, wet lint bags in order to empty them. At times, plastic linen bags break while transporting them from the medical floors to the laundry facility. Furthermore, the bags are often overfull and not tied shut; thus workers are forced to pick up heavy, soiled linens off the floor. Not only are they at risk for back injury, they are also exposed to disease.

1.3.3 Reaching and Forceful Exertions

Large laundry bins have deep bottoms, which may cause personnel to bend in unsafe ways as they attempt to transfer the linens from the bins to the washers. At times, when linens get tangled within the large washers, workers have been known to climb into washers to remove tangled wet linen.

Photograph 1.2 Employee bending to remove linen from deep bin.

1.3.4 Floors

1.3.4.1 Mopping

Dumping water from large mopping buckets into floor drains is often difficult; it may require lifting the bucket over a raised lip around the drain, or lifting to tilt into a sink.

Leather work-boots are slippery when working on wet floors. Without rubber-soled shoes, workers may have to hold onto a buffer with one hand, and the rail on the wall with the other hand for balance. Due to repetitive motions, mopping causes pain to back and wrists.

1.3.4.2 Signage

Healthcare staff may ignore or fail to notice hospital signage, such as "Wet Floor" signs. Even if they notice these signs, however, they may be poorly constructed and may confuse workers regarding the appropriate action to take. For example, workers may not know whether they are to walk on the right side or left side of the sign. Of course, someone who has just mopped the floor may forget to post the signs altogether.

1.4 WHAT SPECIFIC STEPS CAN HOSPITALS TAKE TO REDUCE THE RATE OF INJURIES SUSTAINED BY HEALTHCARE WORKERS?

Vredenburgh (1998) recently found that a hospital's overall management style is a good predictor of its employees' injury rate. For example, hospitals with the lowest injury rates are also those most likely to focus on prevention, rather than on remediation after injuries have already occurred. Preventive steps include verifying that employees are using the skills they learned in ergonomics/safety training and showing hiring preferences for applicants that have a strong health and safety employment history (based on the premise that the best predictor of future performance is past performance).

It is also important to ensure that the hospital's risk manager is included as part of the hospital's management team. Because of the high turnover rates that characterize hospital settings, new employee selection is an ongoing process. As a result, this offers a significant opportunity to select employees based, at least in part, on their prior safety records.

It is worth noting that hospital risk managers have told me that their job candidates' previous employers are not usually willing to disclose their injury records. According to Huck, an employment consultant at Human Resources International, one way to obtain this information is through a behaviorally-based interview. The following questions may be used. In each case, the applicant should be asked to describe the situation, the action he/she took, and the result:

1. In your past experience (work or school), have you noticed any process or task that was being performed unsafely?
2. Everyone has to bend some safety rules at times to get a job done. Please share some examples of when you had to do this.
3. Describe a time when your supervisor, either formally or informally, had to talk to you about a safety violation.
4. Please provide an example of when you had to call a co-worker's attention to a possible violation of a safety regulation.
5. Please describe the types of accidents, or near misses you have had in your current or previous jobs.

In addressing the specific WMSD injuries discussed in the previous section, there are several ways for hospitals to reduce employee injuries. Some are management controls (changes in policy), some involve training employees in proper body posture, while others discuss the use of different equipment or tools to facilitate job performance.

1.4.1 General

In order for a safety and ergonomics program to be successful, it is critical to instate a policy that conveys to employees that safety is not merely a priority, but a value. The policy should tell them, in no uncertain terms, that they will never be asked to compromise their safety. That means that they will never be forced to rush while at work. The policy should also state explicitly that employees from one department will not rush employees from another (i.e. nursing hurrying housekeeping). Finally, encouraging a team approach when a quick change-out of highly soiled rooms is required, may also ensure employee safety.

1.4.2 Lifting and Moving Heavy Loads

1.4.2.1 Lifting and moving patients

- Implement a program to increase the number of employees trained and proficient in proper lifting techniques in order to assure that enough qualified personnel are available for lifting of patients. A pay incentive may be offered to trained/certified lifters.
- Initiate team effort between nursing staff and housekeeping staff to assist in lifting, in order to ensure that sufficient employees are available for difficult and heavy lifts.
- Provide education and training to all employees whose job may involve patient handling regarding availability and appropriate use of lifting aids, mechanical and motorized lifts.
- Post "Fall Precaution" signs outside of the rooms of patients who may be unsteady or confused if they attempted to leave their beds unassisted.
- Develop a checklist to identify patients who qualify as a "fall precaution."
- Provide training in how to control the fall of a patient—have patient slide down your body and legs. Gait belts can be used when walking with a patient to steady them at their center of gravity. Require the use of gait belts on patients who have been identified as "fall precaution."
- Establish guidelines regarding how many people should be used for lifting patients of varying weights. Lift codes can be used effectively in many hospitals.
- A physical therapy evaluation could indicate whether patients (especially heavy ones) have enough upper body strength to assist with lifting themselves.
- When moving patients gurney to gurney, draw sheets are typically used. Garbage bags can help slide heavier patients easier. Plastic slide boards may be an option. Train employees in use of slide boards and similar devices.
- Make sure employees are aware of the capabilities and limitations of the available mechanical and motorized lifts.

1.4.2.2 Furniture, beds, equipment and supplies

- Instate a policy that furniture shall not be moved until it has been emptied.
- In storage rooms, do not stack heavy supplies more than 5' high.
- Investigate the feasibility of developing a "lift codes" program (on-call lifting human resources that are trained, skilled and certified as lifters) to assure that enough qualified personnel are available to lift furniture and heavy objects. A pay incentive may be offered to trained/certified lifters.
- Organize a schedule for moving furniture in order that there are enough hands ready when needed.
- If a housekeeping or facilities manager feels it is unsafe to move an item, he/she should have the authority to hire a professional mover.
- Align, oil, and test wheels of all beds and gurneys, to make sure they all roll together.
- Replace old beds that do not roll (or place in locations where they will not have to be moved).
- Repair and maintain old beds until they can be replaced (brakes, lifting mechanism under mattress).
- Label difficult to move beds as a "2-person move" where they are still in use. Require that two (2) people push beds up/down sloped walkways.

1.4.2.3 Trash and laundry management

- Where there is a suction or removal problem between plastic bags and cans, investigate the feasibility of replacing garbage cans with frames (like laundry) or another can design (such as a coated wire mesh can) to eliminate suction.
- Investigate the feasibility of using smaller, thicker laundry bags (possibly with ties), which will fit on existing frames.
- For orthopedic surgery cases or where there is a lot of laundry, prepare the room with two empty laundry bags. Have surgery team push the first full bag frame to the door to be changed by housekeeping (repeat as necessary).
- Train laundry and housekeeping personnel in proper posture for high to low reach using their legs, not their backs.

1.4.3 Floor Care

1.4.3.1 Signage

Because healthcare workers are often in a hurry, they walk at a fast pace. Therefore, the probability of them slipping on wet floors increases. When there is doubt concerning which part of the floor is wet, put arrows on the wet floor signs to indicate which side of the sign is wet. Make sure signs are installed before floor is mopped and remain until they are dry. Do not wet-mop floors while patients are in their rooms (only to clean spills). Make sure floors are dried immediately.

1.4.3.2 Mopping

Because work-boots are slippery when working on wet floors, require rubber-soled shoes for mopping, stripping and buffing floors (typical requirement is only for closed toe shoes). Train staff in the advantages and disadvantages of both mop-head sizes: a big head is faster with fewer motions but requires more force; smaller heads require more repetition, but less force. Make all sizes of mop heads available to housekeepers in all departments. Mop floors with a dry mop.

1.4.4 Reaching and Forceful Exertions

Train personnel in proper posture for high to low reach using their legs, not their backs. All large bins should have rising spring bottoms so laundry or trash bags are raised to a safe reaching height as the bins are emptied. Provide employees with a long-handled reaching instrument so they do not have to reach or climb into a washing machine to pull wet linen out.

Although healthcare workers are exposed to many work-related musculoskeletal disorders, there are many tools available to help hospitals reduce injuries to their employees. Going back to the initial story at the cafeteria of the Green Forest Medical Center, these employees would have fewer complaints if Green Valley implemented ergonomics controls as part of their employee health programs.

Photograph 1.3 Employees should be trained in the advantages and disadvantages of different sizes of mop-heads

Cal/OSHA, 1997, *A back injury prevention guide for health care providers.* State of California industrial relations.

National Safety Council, 1997, *Accident Facts*, Itasca: Il.

Vredenburgh, A.G. 1998, *Safety management: Which organizational factors predict hospital employee injury rates?* Doctoral dissertation. California School of Professional Psychology, San Diego, CA.

CHAPTER EIGHTEEN

Ergonomic Solutions in Electric Energy Generation

Thomas G. Barracca

Manager, Research & Development, KEYSPAN ENERGY

ABSTRACT

An initial ergonomic evaluation of power plant maintenance activities yielded several beneficial recommendations. Key solutions involved selecting and testing personal protective equipment, analysis of work methods, and engineered solutions for material and equipment handling. The groundbreaking work performed during the first phase of the project was particularly challenging due to the constraints associated with the physical nature of a large power generating station.

One important lesson learned was not to make assumptions regarding off-the-shelf "ergonomic" products. For example, in selecting the best anti-vibration glove for power plant maintenance workers, we could not assume that a glove that had been tested and selected for other utility workers was in turn optimal for plant workers. Secondly, the rapid emergence of many so-called "ergonomic gloves" on the market required KeySpan Energy to perform its own independent evaluation and testing before making recommendations.

By implementing portable adjustable height lift tables, a number of ergonomic challenges were addressed. Since the plant mechanics are constantly on the move, they would often manhandle parts and equipment. Compounding the problem was that there are often no defined workstations throughout the power plant, and mechanics often were working on the floor. The lift tables provide a customized workstation, which allow for improved ergonomic work methods.

An important engineered solution under development is the portable tool balancer. While performing turbine overhauls, manually operated heavy pneumatic tools are extensively utilized. Due to the constraints of the overhead crane system and limited floor space, balancers cannot be fix-mounted to the ceiling or other structures. Portable jib cranes and articulated arms are being explored to address this challenge.

ANTI-VIBRATION GLOVE ASSESSMENT AND RECOMMENDATIONS

During the first phase of the project, vibration was identified as a significant contributor to added ergonomic stress to workers for many power plant maintenance activities. The project team first believed the course of action with recommending anti-vibration gloves would be relatively straightforward. KeySpan Energy had evaluated anti-vibration gloves as a part of an ergonomics research and development (R&D) initiative for our gas construction and maintenance field workers several years ago. At that time, a leather work glove with an anti-vibration palm pad provided excellent vibration dampening with the fit and sturdiness required for gas work. Two factors that caused us to reevaluate the gloves were the multitude of new products which recently came on the market, and more importantly, we did not automatically assume that findings were relevant from one operation to another.

The project team conducted a product search and subsequently screened a majority of gloves based upon user requirements. For example, inflatable air cushioned gloves, although having the potential to provide superior vibration reduction, presented a high probability of deflation due to puncture and relied heavily on operator intervention to properly inflate them. The screening process led us to the detailed evaluation of three different anti-vibration gloves.

The gloves tested included the leather anti-vibration gloves (Impacto) used by gas operations as well as two Decade products (All Purpose and Shock Impact models). The vibration testing was performed using a ¾" impact gun and a tri-axial accelerometer configuration mounted between the padding of the glove and the hand using ANSI standard S3.34-1986. All three gloves were benchmarked versus the standard issue work gloves and all offered significant vibration reduction. The All Purpose vibration-reducing glove rated the best, reducing vibration twofold as compared with the standard issue work glove. In addition, the selected glove had a reduced incremental cost versus the other two models tested. Just as importantly, subjective survey data of plant maintenance workers using the alternatives preferred the fit, feel and durability of the All Purpose models.

KNEE PROTECTION ASSESSMENT

The overall ergonomic assessment observed a number of plant maintenance tasks being performed in a kneeling position. Where kneeling was not the preferred ergonomic posture for performing certain tasks, work methods and equipment changes were evaluated (reference section on "Equipment and Material Handling Solutions"). Due to space and the nature of the power stations, some kneeling is unavoidable. Initially, many workers did not utilize the standard issue kneepads mainly due to issues with ease of use, comfort and perceived impedance of work performance. With this in mind, our major focus was the objective side of knee protection. In other words, the best knee protection would be one that the greatest number of workers would use. Both knee pads and kneeling mats were evaluated and survey data was recorded. The results were spilt between one type of pad and one mat. With Solomon-type wisdom, the recommendation was to provide both kneepads and kneeling mats to increase expected knee protection utilization to close to 100%.

HAMMER TIME

A number of different types of hammers are used in the power plants in a variety of manners. The standard issue wood handled hammers were tested objectively for impact shock versus equivalent composite handle hammers from Nupla. The composite handle hammers reduced peak impact vibration from a range of 16% to 76%. Coupled with positive employee feedback on the composite hammers, recommendations were made to switch.

One interesting note was in the area of sledgehammer evaluation. The composite handle sledges were greatly preferred by plant workers demonstrating them. However, initial laboratory measurements did not show reduced vibration levels for the composite handle sledgehammers. When presented with these findings, the mechanics questioned the tests performed by the ergonomist. Sure enough, the shock impact tests were rerun measuring vibration when hitting a striker wrench, and the composite handle had 55-60% less vibration transmitted.

EQUIPMENT AND MATERIAL HANDLING SOLUTIONS

Due to the nature of the power plant environment, getting equipment and materials into position is particularly challenging. Personnel performing the overhaul work generally have to go into the work area and work around large immobile objects. Space constraints usually do not allow for workstations to be established near where the work needs to be performed. Overhead cranes and side rail gantry systems are mostly geared toward moving very heavy objects. Since maintenance workers are continually moving and are often working independently traditional conveyors, fixed point material handlers or forklifts are not a good solution.

Observation of mechanics performing turbine work found that parts would often be removed from the turbine and then would be off-loaded to the surrounding turbine deck floor or to a traditional cart. Either scenario involved excessive bending to place the part or equipment to a level below the knees. If the part were to be transported via the cart, the mechanic would again need to stoop down to off-load the part at another location. Often, the mechanic would perform maintenance activities (clean, weld, grease) adjacent to where the part was removed. This would frequently involve working off the floor on hands and knees or in a bent-over position. Whether moving or working on parts removed from the turbine, the mechanics were not in an advantageous ergonomic posture.

Based on the needs of plant maintenance workers and the constraints of the power plant environment, our focus was to evaluate the suitability, preference, performance and ergonomic benefits of portable scissor lifts and adjustable height carts/tables. After screening many commercially available products as well as purchasing several for demonstration, for the purpose of this type of work, Autoquip Corporation's Tiny Titan cart appeared to meet most criteria. This cart possesses a 20 X 39-inch tabletop, a raised height of 39 inches and a lifting capacity of 1,100 pounds. The cart received high marks from plant workers surveyed; more importantly, the cart often "disappeared" in the plant. This is a sure sign that mechanics were borrowing it, and it was in demand.

Despite the strong subjective feedback, KeySpan Energy's ergonomic consultant, Ergonomic Technologies Corporation, performed a quantitative assessment of the ergonomic benefits of improved work methods associated with the portable lift table. The assessment was conducted using one worker, a mechanic with 26 years of work experience, for the following test conditions:

1. Inspecting a 4-inch water line valve under normal conditions (e.g., working on the ground).
2. Inspecting a 4-inch water line valve while using the cart as a work surface.

Data collection consisted of videotaping the work being performed as well as the use of a Lumbar Motion Monitor (LMM) for both test conditions. The LMM is an objective measurement device used to quantify three-dimensional movement of the back to identify and/or prioritize manual material handling tasks that increase the risk of occupational lower back disorders. This exso-skeleton was worn by the subject while completing the task to collect real-time postural data (at 60 Hz) including flexion, twisting and side bending. Velocity and acceleration of the deviations were also collected. The LMM and illustrations of back postural deviations are shown in the following figures:

Inspection Using Current Method   Inspection Utilizing Adjustable Cart

1. For this task, inspecting 4-inch water line valves, the Tiny Titan cart provided significant benefits in reducing awkward back postures, including average forward bending (81% reduction).
2. Dynamic back motion, in terms of velocity and acceleration, during the task were low when working with or without the cart.

3. Since heavy lifting was not a significant factor of this task, although handling the valve does occur, it is appropriate to assume that the risk of injury is significantly reduced with the reduction in back bending.
4. The Tiny Titan cart reduced the task duration from 25 minutes to 15 minutes (e.g., 38% reduction). Task or work efficiency is almost doubled with the use of the cart.
5. The total cost of purchasing <u>nine</u> Tiny Titan carts would be justified through the prevention of <u>one</u> back injury.
6. Since kneeling on hard surfaces (e.g., cement floor) is eliminated when using the cart, exposure to high contact forces on the knee and the subsequent risk of injury is also eliminated.
7. In addition to using this cart as a work surface, ergonomic benefits would also be realized by using the cart to transfer the valve throughout the plant. This would limit the lifting or handling exposure of the valve. Small toolboxes could be placed on the cart as well.

Even though the results for this cart were excellent, some plant workers required additional lifting capacity with more vertical travel and a larger tabletop. A double scissor lift table from Vestil with a 2,000-pound capacity and a raised height of 60 inches was also procured.

TOOL BALANCERS

Ergonomic challenges are often encountered in the power plant enabling mechanics to position and operate tools. A typical example would be using a HYTORC drive tool that can weigh nearly 50 pounds. This type of equipment is regularly moved and used at numerous locations. The proposed solution was to utilize a tool balancer to reduce the ergonomic stress caused by lifting and positioning the equipment. The engineering challenge was to make a fully portable platform for the balancer. This was compounded by a narrow aisle situation at KeySpan Energy's older power plants that would typically allow up to a 3-foot wide platform. Therefore, forklift accessibility was not possible to transport the rig or provide counterbalance weight. Two alternative approaches needed to be explored in order to provide coverage for both KeySpan Energy's older/smaller plants and its largest plant that had better access.

In attempting to secure bid specifications for the development of a custom-engineered solution, KeySpan Energy encountered several false starts. Initial bids for portable systems to mount jib cranes with up to a 16-foot arm (reach) or a smaller 10-foot articulated arm raised many questions as to the feasibility of the concept given the high price estimates to build the system. Since the performance and user friendliness of the portable balancers was uncertain, an alternate approach was pursued. A smaller scale 8-foot jib crane was leased and retrofitted to provide counterbalance/portability in order to provide an alpha unit for a basic demonstration.

Utilizing the Portable Jib Crane Prototype

Making Ergonomic Changes in Construction: Worksite Training and Task Interventions

Steven Hecker, MSPH, Billy Gibbons, MBA and
Anthony Barsotti, CSP

1. INTRODUCTION

Workers in the construction industry suffer back injuries and sprains and strains involving lost workdays at rates second only to transportation workers, according to Bureau of Labor Statistics data (CPWR, 1998). Surveys of construction workers have found high prevalence of musculoskeletal symptoms (Hölmstrom et al. 1995, Cook *et al.* 1996), and some clinical studies have supported the link between construction work and musculoskeletal conditions (Hölmstrom *et al.*, 1995). Task evaluations of construction sites reveal numerous ergonomic risk factors common to a number of trades, including awkward and static postures, excessive force and vibration, repetition, and extreme environments (Schneider and Susi, 1994). Other studies have revealed trade-specific ergonomic risks (Spielholz *et al.*, 1998, Zimmerman *et al.*, 1997). The nature of construction work and characteristics of the construction industry create greater obstacles to implementing ergonomic improvements than in fixed-site industries, including:

- A work environment that is constantly changing.
- A mobile work force and short duration jobs that limit employer incentives to invest in prevention of chronic conditions. Even where interest is present, effectiveness of interventions is difficult to measure.
- Inadequate communication among contractors and subcontractors on multi-employer worksites.
- The location of many construction tasks at floor or ceiling levels.
- The common practice of workers supplying their own hand tools, so tool interventions must take place through individual workers rather than through employers. (Schneider, 1995).

This chapter describes an ergonomic intervention research and demonstration project which began in 1995 on the building site of a large semiconductor research and manufacturing facility in the Pacific Northwest U.S. Our research aimed to identify ergonomic risks encountered by trades on this site, develop and implement interventions to reduce those risks, and evaluate the effectiveness of those interventions. This chapter first describes the action research approach we used in developing both programmatic and task-specific ergonomic interventions on this site. We then present some of the intervention outcomes from the initial phase of the project (1995-96). The remainder of the chapter analyzes the effects of a site-wide ergonomic training program that was implemented based on findings from the initial research. We conclude with some lessons and recommendations for ergonomic application to the construction industry.

2. SITE DESCRIPTION AND METHODS

A university team carried out this research in conjunction with an Oregon-based general contractor. This company is engaged in contracting and construction management activity on a variety of industrial, commercial, and government projects regionally, nationally, and internationally. The research took place during two distinct periods of construction between 1995 and 1999. In the initial phase, 1995-96 and referred to as phase one, the first production facilities were built, consisting of a central utility building and clean room "fab" for semiconductor manufacturing. Phase two, 1998-99, involved expansion work within the original buildings and some new construction. In phase one, the general contractor performed structural and mechanical work through its own subsidiaries, and subcontracted about 60 percent of the work to additional contractors. In the second phase, all work was subcontracted.

During phase one, approximately 3,000 craft workers cycled through the site, with peak employment of about 1,200. The construction workforce was nearly 100 percent unionized. Composition of the trades varied according to the stage of the construction process. The high technology character of the buildings resulted in a dominance of electrical and mechanical trades (electricians, pipefitters, sheet metal) during much of the construction, but carpenters and laborers were also well-represented on the site. The phase two workforce had similar demographics and peaked at about 500.

2.1 Safety and Health Program

During phase one, new hires attended a two-hour orientation that was devoted almost exclusively to safety issues, and foremen went through additional safety-related training. Classes on first aid, hazard communication, fall protection, confined spaces, and cleanroom protocol were provided as appropriate. Ergonomics received very brief mention in the new-hire training, focusing on use of mechanical devices or team lifts for material handling tasks, proper body mechanics, and the cumulative nature of many back injuries. Foreman received additional training related to the pre-work stretching program which was required on the site. This training covered the anatomy of the spine and discs, causes of musculoskeletal injuries, and instruction in proper performance of the stretching routine.

A safety group leader (SGL) system was integral to the site safety and health program. The group leaders were selected from trade foremen and conducted biweekly safety meetings with their crews. During these meetings, information from contractor management and safety staff was shared with the work force, and safety and health concerns were brought forward by the workers. Following these small-group meetings, the group leaders met for one hour with site management and safety personnel to bring forward employee concerns and to solve problems between or among crews and contractors. The number of SGLs varied with the size of the workforce. During the peak of phase one approximately 70 SGLs were active.

The site also had an "Injury-Free Workplace" program (IFW), which was introduced by the general contractor and owner through an outside consulting firm. Most managers and foremen of the general contractor and subcontractors went through some training under this program. A primary goal of the program was to improve communication regarding safety among the owner, the general contractor, subcontractors, and craft employees sharing the site. Another goal was to foster the belief that accidents and injuries are preventable in construction and that an injury-free workplace was a feasible goal. The IFW program talked about this goal in terms of a cultural change or paradigm shift for all construction personnel. A Safety Leadership Team (SLT) implemented and monitored the IFW program. The general contractor's project manager chaired the SLT, and members included managers and superintendents from the owner, another general contractor operating on the campus, and some subcontractors.

2.2 Research Hypotheses, Methods, and Activities

This research and demonstration project was grounded in two hypotheses:

1. Construction trade workers are exposed to significant risk factors for musculoskeletal injuries.

2. Construction workers make frequent adaptations of their tools, equipment, and work practices to conditions of the environment. Some of these adaptations are ergonomic in nature. An intervention process which provided ergonomic awareness training and engaged workers regarding their own ergonomic practices would lead to broader implementation of ergonomic modifications and would result in reduced exposure to risk factors for musculoskeletal injuries.

A participatory action research methodology was chosen for several reasons. Action research involves the close cooperation between researchers and subjects in problem identification, planning, strategic action, and evaluation (Israel *et al.,* 1992)` Any proposed intervention had to be accepted by workers and contractors before it could be implemented and evaluated. Acceptance is more likely where the parties have been involved in planning and development of the specific changes (Israel *et al.,* 1992, Noro and Imada, 1991). The multiple trades and contractors on the site represented many different types of work, tools, and equipment. The workers' and contractors' knowledge of the details of these factors was indispensable in designing interventions. The rapidly changing site required that interventions be implemented on a fast track, while specific work tasks were still being performed. This would have been impossible without close communication between researchers and construction personnel.

Multiple methods and data sources were used to identify and prioritize ergonomic problems and design interventions:

1. Researchers reviewed records of prior injuries and illnesses at the site at the start of the investigation and identified cases involving musculoskeletal disorders or ergonomic risk factors.

2. A literature review identified ergonomic risks associated with specific construction trades and tasks. We examined those pertaining to trades active at the research site for possible interventions.

3. The ergonomist member of the research team performed reviews of every accident at the site which resulted in a musculoskeletal disorder or appeared to involve ergonomic risk factors.

4. We trained foremen of two subcontractors on ergonomic fundamentals. The trainers encouraged foremen to identify elements of their own trade's work that put them at risk for musculoskeletal disorders. Facilitated discussions identified specific changes in tools and equipment that would reduce these risks.

5. Researchers conducted periodic walk-through inspections of the site, observing work operations and interviewing workers and foremen, to identify specific ergonomic problems.

Interventions targeted ergonomic risks and included both task-specific and programmatic activities. These were developed in concert with the general contractor and specialty contractors where appropriate. Interventions included both engineering and administrative controls in the following categories:

- Ergonomic risk factor identification
- Tool and equipment modification
- Work practice modification
- Training
- Prework stretching

In some cases, multiple categories were combined in a single intervention.

2.3 Outcome Measures and Data Sources

Given the limited previous application of ergonomics to construction, we were interested in both process and outcome measures and in programmatic and task-specific changes. The challenges of measuring outcomes in intervention studies, particularly those involving work-related musculoskeletal disorders (WMDs), are well-described by Zwerling *et al.* (1997). Quantitative measures of both WMDs and ergonomic risks pre- and post-intervention, while desirable, were in most cases beyond the scope of this study.

We documented specific ergonomic interventions which resulted from accident reviews or ergonomic task analyses using before and after video, photography, and interviews with workers and supervisors directly involved in the tasks and the development of control measures. These observational and interview methods allowed us to assess ergonomic risk reduction in a qualitative way. We gathered data concerning the overall process of integrating ergonomics into the site safety program through participant (researcher) observation, interviews with key informants, and analysis of documents that emerged from various safety bodies and processes. We also monitored the patterns by which particular contractors, groups of workers, or tasks became intervention targets. We pursued some intervention cases because opportunities were provided by cooperative contractors or workers. Others fell into the category of "snowball" sampling in which one intervention idea led to work with another group on a similar or related idea (Patton, 1990).

A written questionnaire administered to the entire workforce gathered cross-sectional data on the prework stretching program. The questionnaire was developed in conjunction with site personnel and pretested on representative groups of workers. The survey was administered through the safety group leader structure. We used a second self-administered questionnaire to assess the sit-stand intervention with pipefitters. The Iowa musculoskeletal symptoms and job factors survey was administered in June, 1996 for comparison to other construction worker populations, but several items specific to this site were added for this population.

The variety of methods, data sources, and subject populations were intended to provide a degree of triangulation and the ability to assess the generalizability of findings.

3. RESULTS

3.1 Musculoskeletal injury incidence

Review of the site's detailed injury records for the period prior to the start of the research intervention found that work-related musculoskeletal disorders were more common than it first appeared in the contractor's summary statistics. An aggressive return-to-work policy for injured workers placed most injured workers back on the job without lost workdays. These workers were returned to broad classifications of work, so that while they may not have been able to do the same task they were doing at the time of injury, they were recorded as performing the essential duties of that job classification. Some musculoskeletal disorders were treated as first-aid cases and without further investigation the specific ergonomic risk factors involved were not identified, thereby masking the severity and specifics of some cases. Participation of the research team in WMD injury reviews led to much closer examination of ergonomic factors in all work-related injuries and incidents. Ultimately for the entire period of construction 43 percent of all injury claims fell into the musculoskeletal category, and these accounted for 92 percent of total claim costs (source: contractor's injury statistics). This injury figure corresponds closely to the percentage of lost-time injuries which are musculoskeletal in nature (46 percent) in national construction statistics (CPWR, 1998).

3.2 Specific Ergonomic Interventions

Specific interventions resulted from incident/injury reviews, follow-up to ergonomic task analyses carried out by the research team, and training presented to a group of foremen. Three of these are described below:

1. An electrician suffered a hernia as a result of straining to pull heavy electrical cable from a large cable spool. The incident review revealed specific risk factors, including a) high force requirements due to friction from the spool bearings and b) poor access to the cable due to material storage. Recommendations based on the review were presented to the subcontractor. The company applied its own continuous improvement process to the issue, reformulated the recommendations, and implemented changes. These included modifying the rack on which the spool hung so that the spool spun more freely, and rearranging stored materials so that workers could access the spool from directly in front.

2. The concrete floor of the clean room is known as a "waffle deck," consisting of thousands of 15-inch diameter circular metal "pop-outs" occupying the center of each square waffle section. The popouts accommodate the frequent rerouting of utility lines into the cleanroom without necessitating core drilling of the concrete. Because of the stringent sanitary requirements in the finished cleanroom, laborers were required to remove each of these pop-outs, clean out residual material from the rim in which it sits, wipe this rim down with isopropyl alcohol, and reseat the pop-out. They performed this work on hands and knees. Researchers collaborated with the foreman and crew to introduce two controls for this task. All workers received kneepads. A low four-wheeled seat, similar to a mechanic's trolley, was purchased and modified so that the laborers could sit while performing the cleaning task. Workers could then perform the job with less extreme bending of the torso, though it still involved significant low-back flexion. The solution was limited in that such seats were not provided for all workers and

additional controls such as job rotation were not implemented. However, it did improve a task that combined high repetition and awkward and static postures.

3. The research team identified ergonomic risks for pipefitter welders including standing on concrete for entire shifts and welding at awkward heights requiring significant neck and upper back flexion. We suggested anti-fatigue matting for the concrete surfaces and sit-stand stools to relieve the constant standing and to allow adjustment of work height. The largest mechanical subcontractor decided to make these changes. They purchased the matting and supplied it to many work locations throughout the site. We could not conduct a controlled evaluation of the effects of the matting because of the rapid adoption of the mats throughout the site, but the matting was widely used and anecdotal comments by pipefitters were favorable and frequent.

   We made six sit-stand devices available to this subcontractor. Foremen identified welding workstations that posed risks of static and awkward postures and which were likely to remain in use for more than a few weeks. They also designated individual pipefitters to use the stools. Prior to using the sit-stands the users were interviewed concerning their usual work posture and work surface and any pain or discomfort they associated with standing on it. They were asked to complete a one-page survey three times per week at the end of their shift. We initially interviewed twelve pipefitters (the devices were used on two shifts). Seven completed at least one subsequent survey, but only two completed as many as five separate surveys. Because so few employees completed more than one or two follow-up surveys, only a general summary of the responses is provided here.

   Most of the pipefitters spent 8-10 hours per day standing. The most common work surface was concrete covered with vinyl, or occasionally with plywood. Prior to the introduction of sit-stands the main types of seating used were trash cans, buckets, or pipe stands with cushions. Few workers used the sit-stands at all and those that did used them for only 1-2 hours per shift. Subsequent feedback from worker interviews revealed that there is still significant resistance to "sitting on the job" among the trades. Sitting was equated with not working and some claimed that owner personnel reinforced this perspective when they observed construction workers. Some of the pipefitters did modify their workstations by changing the height of their welding equipment so that neck flexion was reduced while performing orbital welding.

These are three among a number of ergonomic interventions proposed. In other cases, attempts to identify specific measures failed or, where we could identify such measures, no effective route of implementation was available. The drywall foremen who attended a 90-minute ergonomics training were receptive to the information provided. A significant ergonomic and safety issue was identified, the carrying and hoisting into place of 12-foot sheets of drywall, especially as the work spaces became tighter as the building became closed in. The ergonomist presented the problem to preshift crew meetings and gave the drywallers an exercise to design a solution to the problem, but no solutions were proposed.

Similarly the screw guns used by drywall installers contributed to high force, repetition, and non-neutral wrist postures. These tools were equipped with a pistol grip, but workers invariably used a power grip on the handle to apply more force, using the fourth or fifth finger on the trigger. A manager indicated that the contractor supplied inexpensive guns of modest quality because they received rough treatment. The company didn't purchase more expensive self-feeding screw guns because it didn't expect them to last long. We learned from a major manufacturer that they were studying a redesigned handle for this tool. However, a D-handle or other power grip was not available at the time of the intervention.

3.3 Training

The ergonomics training provided to all new hires was negligible as an educational intervention, especially as it came with a large quantity of other safety information. Access to employees for more extensive training was very difficult to obtain. The Flex-n-Stretch® training for foremen was a more extensive exposure to the nature of WMDs and their prevention, but the time that would have been required to provide this to all workers was not budgeted. Foremen were expected to pass on this knowledge to their crews as they led them in the stretches.

The 60- and 90-minute ergonomics training provided to pipefitter foremen and drywall foremen, respectively, had more noticeable direct effects. The pipefitter training led to the interventions involving anti-fatigue matting, sit-stand stools, and adjustment of working heights. The matting, especially, was immediately adopted by the subcontractor, and within three days of the training was seen in almost all areas of the fab and utility building where workstations of more than a few days duration were found. Sit-stands were placed more selectively, but again their use grew out of the training.

3.4 Prework Stretching

A detailed evaluation of the Flex-n-Stretch® program is provided elsewhere (Hecker and Gibbons, 1997), but we present a brief summary of relevant findings here. Performance of Flex-n-Stretch® was the most consistent reminder to most workers of the emphasis on prevention of musculoskeletal injuries. In the initial survey, 86 percent of respondents reported doing the exercises at least three days a week (n=685, response rate=71 percent), while in the Iowa survey six months later, 82 percent said they stretched always or most of the time (n=380, response rate=68 percent). These data suggest widespread awareness of the program. Frequency of performance varied by contractor in the initial survey, ranging from 20 to 98 percent. Seventy-eight percent of respondents also reported receiving training in performance of the exercises, and the trained workers were significantly more likely to perform the exercises regularly than those who were not trained. The strongest self-reported effects of doing the exercises were on alertness at the start of the shift, awareness of ergonomic risk factors, and flexibility.

We should note that evidence in the literature on the effectiveness of workplace flexibility and exercise programs in preventing musculoskeletal injuries and symptoms is limited and mixed (Silverstein *et al,*. 1988, Hilyer *et al.,* 1990, Skargren and Öberg, 1996). However, the positive employee perceptions found in our study is consistent with most other studies, and more and more construction contractors seem to be making commitments to various types of stretching programs.

3.5 Soft Tissue Injury Prevention Program

The researchers collaborated with the contractor's safety management team to produce a "Soft Tissue Injury Prevention" policy and program (STIP), which the contractor then adopted for the site. This policy codified much of the ergonomics activity that had gone on during the project.

The STIP policy established four main guidelines for the prevention of WMDs in construction:

1) Eliminate the need for heavy manual materials handling (MMH)
   a. Plan material handling tasks and use mechanical lifts wherever possible
   b. Plan work heights to minimize lifting and bending

2) Decrease the demands of manual materials handling
   a. Plan the work area to minimize travel distance with loads
   b. Change the type of MMH activity, e.g. push rather than pull, lower rather than lift, etc.
   c. Rotate jobs to avoid repetition
3) Minimize stressful body movements through attention to body mechanics
   a. Plan work layout to minimize twisting at waist
   b. Use good lifting techniques

4) Stretching and conditioning
   a. Each worker receives training in proper stretching methods and performs stretching sequence at beginning of every shift.

4. PHASE ONE OUTCOMES

We conducted an assessment of the impact of these ergonomic activities at the conclusion of phase one construction. This section summarizes those findings. The concept of ergonomics as applied to prevention of WMDs was new to the contractor and to the site. While the research team introduced the concept and application of ergonomics, the general contractor's safety director had made the initial contact concerning ergonomic intervention on his projects, and the contractor had already brought the Flex-n-Stretch® program to the sites with owner approval prior to the beginning of the research. This indicates that the contractor recognized that WMDs were a significant problem and was looking for solutions.

Ergonomics achieved significant visibility on this site because at least some key personnel of the general contractor, the owner, and the Safety Leadership Team

1. believed it fit into the "injury free workplace" concept,
2. saw it as a logical addition to a safety and health program that was trying to be in the vanguard of the construction industry, and
3. recognized that WMDs constituted a significant part of the injury problem.

Specific organizational features contributed to the legitimacy of ergonomics as an issue on this building site:

- The site safety and health manager, who was committed to raising the profile of WMD prevention, was accepted as part of the project management team.

- The Safety Leadership Team provided a forum for reaching contractor and subcontractor management and owner personnel on a regular basis with the progress of ergonomic field work. A biweekly site newsletter, distributed to all employees, contained coverage of ergonomic activities in several issues, so the information also went beyond the management level.

- The researchers contributed to other safety and health activities on site, for example assistance to the SLT in developing a survey instrument to evaluate its own activities and the ergonomist's involvement in most incident reviews.

However, this phase of research also identified two important and related factors that limited the impact of these activities:

1. The inability to reach the workforce at large with specific information about ergonomic risks and what could be done to abate them.

2. The contractors equated ergonomics with "field fixes," those tool and work practice changes that could be made at the worker and foreman level. For the most part they did not achieve an understanding of what was necessary at the management and organizational level to support ergonomic change.

Access to foremen and superintendents, while always subject to time constraints, was possible, but release time from production for craft employees to receive training in ergonomics was more difficult.   Most specific ergonomic task interventions involved small groups of workers.   In these cases the behavior of at least some of those directly involved changed as a result of the interventions.   But other subcontractors and their workers had little direct contact with the ergonomics part of the program.   Most subcontractors allowed researchers to observe work operations and interview employees.   Some became involved through incident reviews of musculoskeletal injuries, training, and trials of new tools or equipment, such as the anti-fatigue matting and sit-stands.   Others cooperated in a general way but did not agree to specific training or intervention activity.   Reasons cited included the cost in time away from production, the unwillingness to share some of their own work practices which they saw as a competitive advantage, and skepticism that anything could be done about certain tasks, even though the ergonomic risks were evident.

We also learned that different problems required intervention at multiple levels in the construction industry.   Even if contractors developed a fuller understanding of ergonomic issues, owners, architects and designers, and manufacturers of tools and materials needed to be brought into the process of identifying and developing solutions.

The researchers and the general contractor's safety team concluded that reaching the entire craft population and supervision in a systematic way on subsequent projects would be one key to expanding ergonomic awareness and activity among employees and contractors.   The general contractor agreed in principle, but an appropriate site and project had to be identified.   Ultimately an expansion project on the same campus provided the opportunity for implementing an ergonomics training element.   We began developing an ergonomics training module, and approximately twenty months after completion of the phase one, we implemented the program described in the following section.

5. PHASE TWO: ERGONOMICS TRAINING INTERVENTION

5.1 Methods

The research site was an expansion, or "build-out" of the semiconductor fabrication facility constructed in phase one.   The work included some structural concrete, but mainly electrical and mechanical systems and interior partitions, floors, and ceilings, many in cleanrooms.   Major trades involved included carpenters, laborers, electricians, pipefitters, and sheet metal workers.

The ergonomics training program was presented biweekly to employees.   The intent was that employees attend shortly after their arrival at the site, though this was not always possible when the workforce was increasing steeply.   A key element in planning the intervention was that the general contractor made subcontractors aware in bid negotiations that attendance at the ergonomics training was mandatory for all employees, from craft workers to program managers, and that they should incorporate the time required into bids. To recruit for training sessions the general contractor notified subcontractors to sign up their employees and monitored attendance weekly, following up with those subcontractors who hadn't yet sent their employees to classes.

The content of the training reflected the three complementary prevention strategies that earlier research had identified and that were specified in the Soft Tissue Injury Prevention policy:

1. Body conditioning and stretching
2. Proper body mechanics in performing work tasks
3. Identification of ergonomic risk factors and solutions.

The prework stretching program had become a fixture on this campus, and training for this program was incorporated into the overall two-hour ergonomics session. The specific objectives of the training were that craft workers and foremen would:

1. Be able to identify risk factors for musculoskeletal disorders in their jobs
2. Understand the three-pronged approach to preventing WMDs, including stretching/conditioning, proper body mechanics, and ergonomic modifications
3. Correctly perform the Flex-n-Stretch© exercise sequence
4. Be able to distinguish between good and poor body mechanics for performing specific tasks
5. Be able to look at their own tasks, identify one or more ergonomic risk factors, and recommend and/or implement solutions.

While the objectives are interrelated, this chapter focuses on those aspects of the training specific to ergonomic hazard recognition and change.

The training was designed to fit in a two-hour time period. The sequence of the session is outlined below with sections I - IV presented in the first hour:

I. Purpose of the training
   - Scope of the WMD problem
   - Three-pronged strategy for preventing WMDs

II. How musculoskeletal disorders affect the body
   - Physiology of the spine, muscles, joints

III. Prework stretching
   - Why and how the exercises are performed
   - Practice of each stretch

IV. Body mechanics
   - Applying good body mechanics to work tasks
   - Instruction and illustration of lifting techniques

V. Applying ergonomics to construction tasks

- Ergonomic model of looking at WMDs-"Body mapping": where does it hurt, why does it hurt, what can we change to reduce the risk?
- Ergonomic risk factors in construction tasks
- Examples of solutions: tools, work practice modifications
- Where and how does change take place- multiple organizational levels of ergonomic intervention
- How we want ergonomic change to occur on this site

VI. Evaluation

5.1.1 Training Methodology

The training was designed in accordance with the principles of empowerment education including:

1. Learner-centered education
   - Content would flow from workers' own experiences
   - Audio-visual and verbal examples would come from the site at which they were working or similar, easily recognizable sites

2. Participation
   - Activities and exercises to directly involve employees
   - Eliciting of employees' own experiences for both problems and solutions

3. Training for change
   - Identification of barriers to change
   - Recognition of multiple levels at which change must take place.

Workers' experience was tapped through the body-mapping exercise, which resembles a simplified symptom survey. Workers were given two dots of different colors to stick on a large human figure drawn on a flipchart to indicate the parts of their body where they experienced the most pain and/or discomfort related to their work tasks. The pattern of dots in each class was used to trigger discussions of what particular tasks contributed to the problem areas identified. As workers described activities and the particular body parts or risks involved, the instructor filled in a list of ergonomic risk factors with the more technical terms an ergonomist would use. These risk factors included those typically used in general industry and others more specific to construction:

- Force
- Static posture
- Awkward posture
- Contact stress
- Repetition

- Vibration
- Environment- hot or cold
- Housekeeping
- Lack of pretask planning

The instructor was the ergonomist member of the research team. She was prepared with videotape examples of many of these risk factors from her prior work on this campus and similar high-technology construction sites. These examples were used to demonstrate problems and possible solutions. The instructor sought other examples of ergonomic modifications from class members as well.

Other participatory elements of the program included practice of the stretches and an activity that introduced the ergonomics part of the program. In this activity volunteers from the group were presented with pieces of paper numbered 1 to 10 arranged randomly on the floor and asked to place them in order in whatever manner they chose. This triggered discussions of the fact that construction tasks are often performed at floor level, sometimes because that is where the installation is taking place but often just because that is where the parts are found.

5.1.2 Evaluation

Three main tools were used to evaluate the effects of the training program. Each participant was asked to complete a questionnaire immediately following the training session. These questionnaires (T1) were distributed in the classroom at the conclusion of the training and collected by the instructor. This survey collected several kinds of data including

- How much new information they received related to stretching/ conditioning, body mechanics, and ergonomics
- Perception of ergonomic risks in their trade
- Perception of the practicality of ergonomic changes and prediction of their making changes
- Personal experience with work-related musculoskeletal pain and discomfort and rating of job factors that contribute to such pain and discomfort.

A second follow-up questionnaire (T2) was administered during the seventh month of the phase two project, five months after the training was first offered. The existing safety group leader communication system was used for distribution of the questionnaires. Prior to the day of the crew meetings each group leader received sufficient questionnaires for his crew with instructions on handing out and collecting completed questionnaires. The group leaders then returned the questionnaires to the safety office. The surveyed population consisted of the construction workers who were on the job during that week. Some of the questions were identical to those on the initial training questionnaire, but new questions were added. Within the four-level evaluation framework developed by

Kirkpatrick of *reaction, learning, behavior, and results*, this survey was targeted at the behavior and results levels (Kirkpatrick 1994). The data collected included:

- Whether and how the respondent applied information from the training on the job
- Perception of ergonomic risk
- Perception of ergonomic responsibility of various parties on the site
- Participation in and effects of the stretching program.

Data from each questionnaire were analyzed separately and the two data sets were also compared to assess change over time that might be attributable to training.

To supplement questionnaire data, researchers conducted periodic observation of work tasks on the site and assessed specific cases of ergonomic change which individual workers and crews brought to their attention. One member of the research team regularly attended the biweekly safety group leader meetings to hear about ergonomic issues and to provide consultation to the group leaders.

6. RESULTS

6.1 Response Rates and Study Populations

A total of 583 individuals attended the ergonomics training session over a 10-month period. Post-training questionnaires (T1) were completed by 479 attendees (82.2 percent). Of these 58 were identified as other than construction workers (office workers, managers and administrators, engineers, etc.) and 14 did not provide their occupation. The findings from this survey reported in the remainder of the chapter are for the 407 respondents who identified as construction workers as shown in Table 1.

The follow-up survey (T2) was administered through the safety group leaders. The population of craft and supervisory workers on the day of administration was 256, with 46 office staff also on site. The questionnaire was completed by 208 respondents for an overall response rate of 68.8 percent. Of these 208, 183 were construction personnel. Calculated in this way the response rate of construction personnel was 71.4 percent (183/256). Table 1 summarizes the responses of all construction workers and the major trades represented. Of the 183 construction worker respondents, 110 had attended the ergonomics training prior to completing the questionnaire.

6.2 Questionnaire Data

Workers at T1 were asked how much of the information in the training was new to them. More than 62 percent of respondents found the information in all three categories conditioning and stretching, body mechanics, and ergonomics to be mostly or all new (Figure 1). The ergonomic information was rated all new by a higher percentage (15.5 percent) than the conditioning and body mechanics information (8.1 and 9.3 respectively).

As a measure of risk perception workers in both surveys were asked how often workers in their trade face ergonomic risk factors such as forceful exertions and awkward postures. Eighty-seven percent of the post-training group felt they faced such risk factors all or most of the time (Figure 2). This figure dropped to 70 percent in the follow-up survey. In both cases less than two percent felt that risk factors were rarely or never present. Variation by trade in these perceptions was slight.

Table 1 Construction worker survey respondents

	T1 Post-training	T2 Follow-up
Total respondents	407 (100)	183 (100)
Trade/Occupation	percent	percent
Plumber/pipefitter	26.8	19.7
Carpenter	18.8	22.4
Electrician	23.1	37.2
Laborer	9.6	3.8
Sheet Metal	14.7	10.9
Other construction	6.4	6.0
Job status		
Apprentice	20.9	24.7
Journeyman	45.0	46.7
Foremen	17.0	21.4
General foremen	5.4	1.6
Superintendent	4.7	2.2
Other	7.1	3.3
Gender		
Male	93.1	90.7
Female	6.9	9.3
	Mean/SD	Mean/SD
Years in trade	14.1/10.8	13.9/10.7
Age	38.8/11.0	38.9/10.9
Musculoskeletal pain/discomfort	Yes	No
Related to work	77.5	22.5
Prevented from days work	18.1	81.9
Seen physician	26.2	73.8

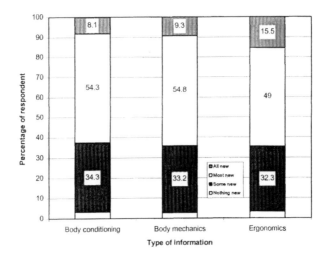

Figure 1 How much new information did you learn?

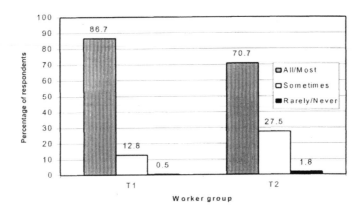

Figure 2 Frequency of ergonomic risk factors in your work

The T1 construction worker population reported on musculoskeletal pain and discomfort which they felt was related to their work (Figure 3). More than 77 percent reported such pain or discomfort within the past 12 months, while 26 percent had seen a physician for this problem during that period and 18 percent had missed work.

Data were tabulated from the body-mapping exercise. The most common location of pain was the lower back, where 303 dots were placed, followed by the neck with 91, and left or right knee with 81. Other frequently noted locations were upper back (55) and right shoulder (45). For upper extremities-- wrist, elbow, and shoulder-- pain and discomfort was much more common on the right than the left side. Of the 752 dots that were placed, 100 indicated "no pain."

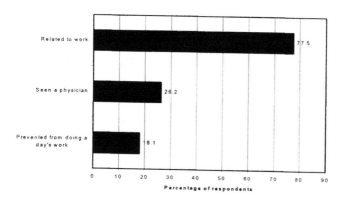

Figure 3 Self-reported work-related musculoskeletal symptoms in past 12 months

Workers at T2 who had attended the training were asked how much of the information they had applied to their jobs in the following areas:

1. Stretching
2. Lifting techniques to avoid straining the body (lifting)
3. Modifying tools or equipment to reduce ergonomic risk factors such as force awkward posture, vibration (tool modification)
4. Changing the way you do your work to avoid ergonomic risk factors (work practices).

Figure 4 shows that for all construction workers stretching and lifting information was applied some or a lot (3 or 4 on a 4-point Likert scale) by more than 90 percent of the workers. Changes in work practices were applied by 70 percent while tool and equipment modification was applied the least but still by more than half (58 percent). These findings were roughly consistent across the four major trades on the site.

Workers were asked whether they had made tool, equipment, or work practice changes as a result of the training. Two thirds of the 110 who had been trained said they had made such changes. Of these 72 respondents who had made changes, 61 provided at least one example. Figure 5 shows the rough breakdown of the kinds of changes made. The largest number of comments were made about various changes in lifting procedures (22), but changes related to working height, work posture and position, appropriate and ergonomically designed tools, and improved planning and design were also mentioned by six or more respondents. This group includes some non-construction personnel.

The T1 questionnaire had asked if the attendees thought they would make ergonomic changes in response to the training. Eighty-two percent of respondents thought they would make changes immediately after training compared with 67 percent who said they actually did in the follow-up survey.

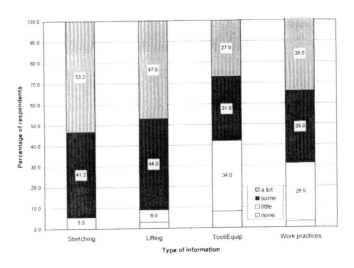

Figure 4 Training information applied to your work

7. DISCUSSION

This analysis of training impact relied upon questionnaire data from two different but related populations. The T1 construction worker population consists of workers who attended the ergonomics training over a period of approximately ten months. While efforts were made to have workers attend the training class within a few weeks of their starting work at the site, this was not always possible, especially during times of steep increases in the work force. Most of the questions in this questionnaire are not likely to have been affected by either duration of work at the site before training or the passage of time over which these questionnaires were completed.

The population responding to the follow-up questionnaire was virtually identical to the T1 population in terms of age and years in the trade, though it differed somewhat in trade make-up. There was small variation in response by trade so this is probably not a significant issue. The follow-up group had higher representation of both apprentices and foremen, but this was largely due to the lesser representation of higher level management because of the way this survey was administered. The purpose of the T1 survey was to evaluate the training itself, assess predictions of how the trainees would use the information, and gather data on musculoskeletal risks and symptoms in this population. The follow-up survey overlaps the T1 survey in some respects, but is primarily directed at what workers actually did with the training. A number of critical questions were asked only of those who had been trained.

Questionnaire responses indicate that most construction workers in the population at this site perceived ergonomic risks to be prevalent in their work. Immediately after the training 87 percent of the T1 group responded that such risks

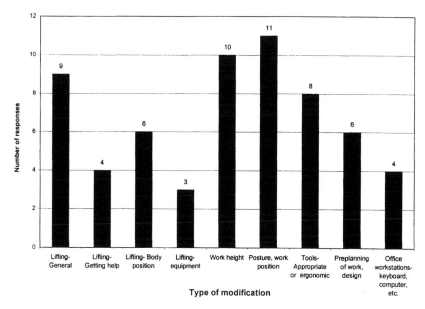

Figure 5 Types of ergonomic modifications made by workers

were present all or most of the time. This figure was lower in the follow-up group but still high (71 percent). The difference can probably be attributed to availability bias from the content of the training and the visuals of construction tasks the respondents had just seen. The perception of high risk was reflected also in the responses on symptoms, where 78 percent had experienced musculoskeletal pain or discomfort in the past 12 months that they felt was related to their work, and 26 percent had seen a physician for this pain or discomfort. The body-mapping results, even accounting for all the subjectivity and peer influence that may have been involved, provides a dramatic visual image of the extent of musculoskeletal pain and discomfort this population experienced.

Much of the material presented in the training was apparently new to this population. The information on making ergonomic changes was slightly more unfamiliar than that on stretching and body mechanics. This is not surprising as 60 percent had participated in stretching programs on other projects, and training on lifting practices is fairly common. Ergonomics training or similar intervention programs, on the contrary, remain extremely rare in the construction industry. Likewise, the respondents were more likely to have applied information on stretching, lifting, and work practices than on tool and equipment modifications. The former are more within the worker's own control than are tool and equipment choices. Those workers who described the changes they had made were most likely to name something to do with lifting (mentioned 22 times). Still, numerous workers did mention ergonomic changes involving principles that were emphasized in the training. For example 10 mentioned moving work off the floor and to waist height. This risk factor was a focus in the training, including the classroom exercise described above, and was reinforced on observational walks by the ergonomist and general contractor personnel. Preplanning and better design of work tasks, and choice of appropriate and ergonomically-designed tools were also specifically noted by six and eight respondents, respectively.

The responses of apprentices differed substantially from those of journeymen in the follow-up survey. The proportion of apprentices who attended the ergonomics training was lower than journeymen, 47 versus 66 percent. Of those apprentices who were trained, fewer applied the various types of information than did journeymen. This was true of stretching (85 applied some or a lot versus 100 percent), lifting (80 versus 98), modification of tools and equipment (35 versus 60) and work practices (50 versus 76). Only 55 percent of apprentices said that the training had caused them to make changes in tools, equipment, or work practices, compared to 76 percent of journeymen. The small number of trained apprentices in the sample (n=20) limits the conclusions we can draw, but several explanations seem plausible. These responses may reflect the lesser control that apprentices have over their work, their fear of doing things differently, and/or a lesser receptivity to the message of making changes to reduce ergonomic risks because they are younger and perceive themselves less subject to strain and sprain injuries. Interestingly, apprentices responded in higher proportion than journeymen that workers in their trade faced ergonomic risks all or most of the time, 90 versus 64 percent.

7.1 Specific Task Interventions

To supplement the questionnaire responses one of the authors did regular observations around the worksite and interacted with crews and contractors concerning work practices and ergonomics. She also regularly attended safety group leader meetings. Craft workers and project supervision who had attended the training session brought to her attention ergonomic problems they had identified. In some cases they were seeking assistance in finding solutions and in others they had already developed their own ergonomic modifications. Several examples are presented:

1. Installation of utilities in a three-foot crawl space. Pipefitters, electricians, and sheet metal workers were installing pipe, conduit and cable, and duct work while working off a 20'x30' concrete platform, 16 feet above the ground. There was only three feet of clearance from the surface of the platform to the ceiling. Because of the height they wore full body harnesses which made maneuvering difficult. Workers had to move materials around either lying on their side or in a squatting position, creating awkward postures and stress on the back, knees, and neck.

   Pipefitters worked with the ergonomist on a task plan to address the ergonomic risk factors. Approval was obtained to have carpenters install guardrails around the platform to eliminate the need for body harnesses. Foam pads were placed in the area so that workers would not be working directly on the concrete. Low-profile material carts were introduced to reduce the need for manually dragging pipe, conduit, and duct material.

2.   Concrete core drilling. A team of four drillers, two above and two below, drilled 3000 12"x6" cores in a concrete deck. The two below positioned and secured steel boxes in place to catch the core and accompanying slurry from the drilling. They had to insert 1/2" all-thread bolts 16" long through a section of strut and thread on a nut to help secure the box. This was done 10-12 times per day by hand, by "rolling" the bolt between the palms. The result was repetitive hand and wrist motions and contact stress at the palms.

   Discussions among the ergonomist, the crew, and the general contractor's safety staff identified a slip-nut that can be slipped on the bolt at any point and twisted to tighten. Thus the all-thread could simply be pushed through the strut rather than screwed for its whole 16" length.

3.   Pipe cutting and reaming. Pipefitters needed to manually cut and ream about 5000 pieces of one-inch PVC pipe with a hand reamer. After attending the training they recognized the repetition and non-neutral wrist posture involved in this operation.

   The crew attached to a drill a PVC connector sized to accept the one-inch pipe. Then they mounted the hand reamer on a pipestand at waist height. They taped a vacuum hose to the back of the reamer to collect the pipe debris, a requirement in this microprocessor facility. To ream the pipe they inserted each length in the connector, pressed it against the mounted reamer, and avoided the repetitive motion of hand reaming.

4.   Threading all-thread rods into floor pedestals. Carpenters were installing a raised floor which involved threading several thousand pieces of 16" all-thread bolts into floor pedestals that later support floor tiles. A coupling was fabricated and positioned onto a drill that would hold the matching head of the all-thread rod. Then the pedestals were lined up at waist height on a rack. The carpenter could then insert a piece of all-thread into the coupling and drill, instead of hand turn, the all-thread onto each pedestal. The power tool eliminated the repetitive hand motion and the rack allowed the work to be performed at waist height instead of bent over or kneeling.

5.   Installation of air filters and blanks. A carpenter crew was responsible for installing a large number of filters and blanks in the ceiling of a cleanroom approximately 12 feet off the floor. The filters weighed about 50 lb. and the blanks 20 lb. Each person would handle 50 to 60 pieces per day. The filters were handed from a worker on the ground up to a worker standing on a rolling scaffold. This worker would then hoist the filter overhead to fit into the ceiling grid. Additional demands were created by space constraints that required moving piles of filters and blanks around on the floor prior to installation. Workers had begun to mark individual filters each time they moved them, and they discovered that by the end of the job many pieces had five or six marks on them.

Just after a crew member suffered an MSD doing this job several other crew members attended the ergonomics training class and subsequently requested that the ergonomist look at this job.

Evaluation of the job confirmed a number of design features that presented significant ergonomic risks, including the weight of the filters and the friction created by the sealant where they were seated in the grid. There was little that could be done to alter the current job, but the identification of this problem resulted in the creation of a "safety in design" task force by the owner. This group was charged with looking at design specifications from the point of view of constructability, both for new construction and retrofitting of the facility.

Adaptations to drills to eliminate repetitive hand/wrist activity were the most common type of modification made on this site. A second common theme was relocating work from the floor to waist height. Most of these ergonomic modifications resulted in improvements in efficiency as well as reducing risk factors.

A small but significant difference can be seen between this set of successful or attempted interventions and those undertaken in phase one. The earlier ergonomic interventions were initiated exclusively by the researchers, although in some cases contractors and workers took them over at some point and carried them out in their own fashion. In the phase two cases most of the changes were initiated by the crews themselves. The training described in this study, modest though it was, seemed to provide a base of knowledge for the worker and manager population that provided a common language and an impetus to identify problems and look for solutions.

7.2 Intervention Matrix

Over the course of our research on this site we have developed a tentative model for the application of ergonomics to construction which locates interventions along two axes: complexity and duration of impact. The four-quadrant matrix in Figure 6 illustrates the current state of the model and the following text explains the thinking behind it.

The X-axis refers to the level of complexity of the intervention. How easily or quickly can it be implemented? How much adjustment is required on the construction site to make this intervention happen and by how many people?

The Y-axis represents how long the intervention is expected to have an impact for the worker. Shorter-term interventions may affect the worker only for the duration of a task or a project. The intervention may not be available for the next task or next project depending on the players or resources available.

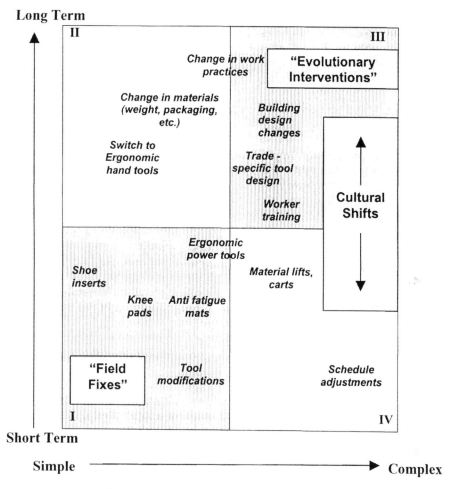

Figure 6 Construction ergonomics intervention matrix

Quadrant I: Interventions in this quadrant are referred to as "field fixes". These are interventions that are identified, performed and implemented at the worksite, usually by the worker. They don't require extensive change in the way work is performed and they require little involvement by players outside the immediate work group. These are "worker interventions" that occur most often because they are easily implemented and are within the worker's span of control. Custom tool modifications, makeshift floor padding, and shoe inserts are examples of field fixes.

Quadrant II: Interventions in this category require little change or effort at the worksite but have a longer impact for workers over time. Ergonomic hand tools don't require too much adjustment on the worksite and can be carried from site to site, usually by the worker. A change in materials may involve the manufacturer making much of the adjustment before delivery to the site. For example, concrete blocks can be produced with handles built in so they are easier to handle on the worksite, or concrete mix can be packaged in 45 instead of 90 lb bags. Others material changes may require more adjustment on the worksite, and are therefore more complex interventions.

Quadrant III: In this quadrant interventions not only have a long-term impact but are more complex and take more time to implement. For example, a building design that does not consider constructability, i.e. how the construction will be executed, is difficult to alter at the time of construction. Communicating ergonomic hazards created by poor design to designers to prevent future construction worker exposures requires an "evolutionary intervention" approach.

Interventions that require significant shifts in the construction culture also fit here. This is mostly due to the mobile nature of the workforce. A cultural approach to intervention does not transfer easily from one site to another. The industry requires repeated exposure to a suggested cultural change in order to absorb it. For example, the acceptability of performing work in a seated position will have to be reinforced again and again at each worksite before workers will overcome years of conditioning to the contrary. If projects are long enough, they can develop their own subculture where a cultural shift can occur just on that job site and/or just for the duration of the project.

Quadrant IV: Interventions in this quadrant are usually dependent on the type of project, the resources available, and the attitudes of the players (i.e., foremen, superintendent, project manager). The introduction of material lifts and carts may depend on whether the foremen know what material handling devices are available or are comfortable making the request to upper management. These interventions usually require involvement at many levels on the project. For example, material-handling equipment may need to be bid into the contract or additional funds approved when the need for the equipment is identified. These modifications may be provided on one site but not on another. This limits the longer-term impact on the workers. Schedule adjustments also fit here because of the high level of involvement and coordination of many site players necessary to make such a change. Scheduling work consecutively instead of concurrently reduces the number of workers in the same work space and provides room for tools and equipment. Addressing schedule pressures on general and subcontractors make this a highly complex intervention.

8. CONCLUSIONS AND RECOMMENDATIONS

This study offered the opportunity to examine the impact of different approaches to implementing ergonomic change in construction over time at a single site. Early efforts focused on introducing ergonomic concepts and recruiting contractors and crews to look at specific tasks that placed workers at risk of WMDs and attempt modifications to reduce those risks. The intervention strategy might be called opportunistic. It produced several specific interventions but also pointed to the need for a more systematic approach to raise awareness about ergonomic risks and possible solutions among workers, contractors, and construction managers and owners. The researchers and general contractor, with owner support, settled on a site-wide ergonomics training module for all construction workers as a means of amplifying the crew-specific effects of the earlier effort. Questionnaire and observational data suggest that this training program was at least partially successful in heightening awareness of ergonomic risk factors and spurring modifications to reduce such risks.

Our data and anecdotal experiences in these cases demonstrate that construction workers and contractors recognize the risks of musculoskeletal disorders in their industry, but how they act on this recognition is subject to multiple forces. The construction community is rooted in traditions and a sense of identity from its accomplishments, including how the work itself is approached. The work is characterized by an unusual combination of conservatism, embodied in the traditional craft tools and work methods, and sometimes radical change as new technologies are introduced, usually with the aim of improving productivity and sometimes quality. Both of these characteristics have strong implications for the application of ergonomics in the construction industry. Ergonomics attempts a systematic approach to the relationship between the worker and the tasks, tools, equipment, and environment of construction. Workers and contractors alike must see some advantage to making changes to these elements of the work in order for measures to reduce ergonomic risk factors to succeed.

The environment in which this research was conducted was unusually supportive of worker safety and health and of innovation. The training provided in phase two reinforced the belief in some contractors and workers that change was possible and desirable. Where crews received substantive support, in the form of time, equipment, and expertise where needed, they made real changes. But diffusion of ergonomics through the industry and even to other sites in the immediate region requires considerable resources and attention to particular issues, including the following:

1. Workers and contractors alike are more dubious about the existence of practical solutions to ergonomic problems than they are about the problems themselves. Just as in fixed industry identification and dissemination of solutions is critical.

2. The structure of employment relationships in construction suggests that ergonomics be integrated into apprenticeship and journey-level training programs. Contractor-by-contractor approaches are welcome but clearly insufficient. In the unionized sector the training centers jointly administered by unions and contractors are a natural vehicle for disseminating this training. Apprentices constitute the most accessible audience for such training, but conflicts arise in that what apprentices are taught in classes is not always accepted practice in the field by journeymen or contractors. One means of resolving this conflict is closer coordination between contractors and training centers to identify the real training needs of workers going into the field. This needs to be

accompanied by journey level workers having a similar understanding of the problems and what can be done about them.

3. Training is necessary but not sufficient to insure ergonomic change. Training needs to be supported with other resources at the owner and contractor level to assist in designing solutions in the field. This is particularly true with smaller sites and contractors, where support of associations and government safety and health agencies may be needed to fill the gap.

4. Scheduling, planning, and sequencing of work have enormous implications for ergonomics on the construction worksite. Many exposures to ergonomic risks have their roots in decisions in these areas, which workers and even contractors often have little control over. Those who make and influence these decisions must be brought into the ergonomics discussion.

5. Individual solutions involving prework stretching and lifting techniques were more readily adopted than design solutions that required changes at multiple levels of the project and industry. But many significant ergonomic improvements will, in fact, require involvement of owners, designers, and manufacturers. Education for these groups is a story beyond the scope of this paper, but one that must not be neglected.

REFERENCES

Center to Protect Workers' Rights (CPWR), 1998, *Chart Book: The U.S. Construction Industry and Its Workers, 2nd edition.* (Washington, D.C. Center to Protect Workers' Rights.

Cook, T.; Rosecrance, J. and Zimmermann, C., 1996, *The University of Iowa Construction Survey*, (Iowa City, IA University of Iowa Biomechanics and Ergonomics Facility).

Hecker, S. and Gibbons, B., 1997, Evaluation of a prework stretching program in the construction industry. In *Proceedings of the 13th Triennial Congress of the International Ergonomics Association, 1997*, (Helsinki: Finnish Institute of Occupational Health), Volume 6, pp. 115-117.

Hilyer, J.C., Brown, K.C., Sirles, A.T. and Peoples, L. 1990, A flexibility intervention to reduce the incidence and severity of joint injuries among municipal firefighters. *Journal of Occupational Medicine*, **32**(7), pp. 631-637.

Hölmstrom, E., Moritz, U. and Engholm, G., 1995, Musculoskeletal disorders in construction workers. *Occupational Medicine: State of the Art Reviews*, **10**(2), pp. 295-312.

Israel, B; Schurman, S. and Hugentobler, M., 1992, Conducting action research: Relationships between organization members and researchers. *Journal of Applied Behavioral Science*, **28**(1), pp. 74-101.

Kirkpatrick, D.L. 1994, *Evaluating Training Programs: The Four Levels*, (San Francisco: Berrett-Koehler).

Noro, K. and Imada, A. S., Eds., 1991, *Participatory Ergonomics*, (London: Taylor & Francis).

Patton, M. Q., 1990, *Qualitative Evaluation and Research Methods* (2nd ed.), (Newbury Park, CA: Sage).

Schneider, S. and Susi, P. 1994, Ergonomics and construction: A review of potential hazards in new construction. *American Industrial Hygiene Association Journal*, **55**(7), pp. 635-649.

Schneider, S., 1995, Implement ergonomic interventions in construction. *Applied. Occupational and . Environmental Hygiene*, **10**(10), pp. 822-824.

Silverstein, B.A., Armstrong, T.J., Longmate, A. and Woody, D., 1988, Can in-plant exercise control musculoskeletal symptoms? *Journal of Occupational Medicine,* **30**(12), pp. 922-927.

Skargren, E. and Öberg, B., 1996, Effects of an exercise program on musculoskeletal symptoms and physical capacity among nursing staff. *Scandinavian Journal of Medicine & Science in Sports*, **6**, pp. 122-130.

Spielholz, P., Wiker, S. and Silverstein, B.A., 1998, An ergonomic characterization of work in concrete form construction. *American Industrial Hygiene Association Journal* **59**(9), pp. 629-635.

Zimmerman, C., Cook, T. and Rosecrance, J., 1997, Work-related musculoskeletal symptoms and injuries among operating engineers: A review and guidelines for improvement. *Applied Occupational and Environmental Hygiene*, **12**(7), pp. 480-484.

Zwerling, C., Daltroy, L.H., Fine, L.J., Johnston, J.J., Melius, J. and Silverstein, B.A., 1997, Design and conduct of occupational injury intervention studies: A review of evaluation strategies. *American Journal of Industrial Medicine*, **32**(2), pp. 164-179.

OFFICE
ENVIRONMENTS

CHAPTER TWENTY

Visual Ergonomics in the Workplace

Jeffrey Anshel, OD

OUR EYES AND VISUAL SYSTEM

Vision is our most precious sense. Our eyes are in constant use every waking minute of every day. The way we use our eyes can determine how well we learn, work and perform throughout our lifetime. Over eighty percent of our learning is mediated through our eyes, indicating the important role our vision plays in our daily activities. The way we use our eyes in our daily routine has changed dramatically over the past number of years. More and more tasks are done at a close viewing distance, and we are working under a variety of workplace conditions. Our visual system must adapt to these changes in order for us to function to our maximum potential.

This paper is designed to give the participants awareness and knowledge of how lighting can and will affect workplace well-being. It also will enlighten the participants to the area of visual function and its role in workplace productivity. By understanding the connection between comfort, health, and productivity and knowing the many options for good ergonomic workplace lighting, readers will learn how to achieve worker-oriented lighting to insure that the task can be easily and productively accomplished. You will also learn how to become sensitive to potential visual stress that can affect all areas of performance.

A complete eye examination is more than just reading letters on a chart twenty feet away. It is simply one test of the function of one part of the visual system. The eyeball is just the receiver of light and the comparison of the eye to a camera is limited in understanding how we really see. Visual processing is accomplished in the brain where visual perception occurs. Eye "sight" is the process of properly focusing the incoming light to the proper area of the retina, whereas "vision" is the process of taking that information into the brain, making "sense" of it and reacting appropriately.

The pathway of light through the eye travels through the cornea, the anterior chamber, the pupil, the lens, the vitreous body and then to the retina, where the light energy is transformed into nerve impulses. This nerve impulse travels out of the eye via the optic nerve, which is simply a mass of nerve fibers, which extend from the retina to the brain. So the eye is actually "attached" to the brain, being an extension of it. When the entire process works normally, the visual state is known as "emmetropia". If the light comes to focus too soon (before striking the retina), it is called "myopia" or near-sightedness. If the light strikes the retina before it has come to a focus, it is called "hyperopia" or far-sightedness. If there is any distortion in the shape of the cornea or other optical structures, then "astigmatism" can occur. This is a common occurrence and often needs an optical correction to compensate for the distortion.

Also important are the fundamentals of binocular vision, with emphasis on the computer viewing environment. The process of coordination between binocular vision and the accommodative or focusing system is presented in a unique forum. Studies have found that the convergence system, where the eyes turn in toward each other as the object moves closer, plays a significant role in vision stress. Additionally, the eyes turn down, as well as in when they view a close object. This results in a normal near viewing posture, which is duplicated optimally with book reading. The viewing of a near object at a raised, or eye-level as is often seen in VDT environments, is awkward and unnatural.

The history of visual demands puts our current-day viewing conditions in a historical perspective. The origin of the visual system and its design shows how our visual system demands have changed. When homo sapien first appeared about 40,000 years ago, we were hunters/gatherers, limiting our time to survival skills. Visually, these included hunting, making weapons, cooking, and little else. Since we are living longer and leading more active lifestyles, the effects of aging on the eye must be addressed. In 1900, the average life expectancy of a male in the US was 47 years old. Now, one hundred years later, it is about 76 years old. We have effectively out-lived many of the useful functions of our eyes. The Bureau of Labor Statistics predicts that there will be a 42% increase in the number of workers over the age of 55 by the year 2002. Aging in the workplace is certainly an issue that needs to be addressed, especially in the area of vision.

COMPUTER VISION SYNDROME

Computer use has grown significantly in the last 40 years. The first computer was developed in about 1950 (when it occupied an entire room), and it is now as commonplace as a telephone in our workplace. There are currently about 70 million Americans using computers regularly in the workplace. That amount is expected to grow to 100 million in the next five years. And now with the growth on the Internet, this projection may need to be revised upward.

The symptoms of physical problems that computer users are experiencing are increasing. The eye care community has also seen a jump in the number of patients who request eye examinations due to symptoms they experience at the computer. This has led to the American Optometric Association (AOA) designation of Computer Vision Syndrome (CVS).

According to the AOA definition, CVS is "the complex of eye and vision problems related to near work which are experienced during or related to computer use." The symptoms that most often accompany this condition are eyestrain, headaches, blurred distance or near vision, dry or red eyes, neck and/or back ache, double vision and light sensitivity. The factors that most often contribute to CVS are a combination of improper workplace conditions, poor work habits and existing refractive errors. Lighting, vision and posture are all inter-related concepts. We are visually directed creatures and will alter our posture to alleviate stress on the eyes.

Therefore, paying attention to body posture may be indicative of a visually stressful situation. Some of the symptoms of CVS actually concern the head, neck and shoulder areas of the body.

WORKPLACE LIGHTING

Lighting is one of the most overlooked and under-emphasized components of our workplace. Whether working at the computer or in a warehouse arena, our field of vision needs to be free of reflections and sources of glare. Our lighting needs to prevent problems, not cause them. Lighting is workplace-effective when it allows the worker to see the details of a given task easily and accurately. Comfort in lighting is a very individual concern and must be addressed on a one-to-one level; no one lighting pattern will work for every working situation. Those in charge of workplace lighting need to learn what is available to help them make the right choices for their employees. Lighting and vision are interdependent factors and must both be considered when designing a working environment for maximum efficiency.

An important factor that affects our ability to see well in the workplace is the quality of light. Good quality light creates good visibility and visual comfort. This can be accomplished with attention to brightness, contrast, quantity and color of light. Contrast between a task object and its immediate background must be sufficient to enable the worker to clearly view the task. Discussion of the contrast ratios will give the attendees enough information so they can set up work areas to maximize productivity without increasing eyestrain. In general, a 2:5:10 ratio is ideal; that is, the task area should be 2-3 times as bright as its immediate surroundings and 5 times as bright as the mid-range area in the office, and 10 times brighter than the peripheral area.

Too much or too little light can inhibit the worker's ability to effectively see the task. Comfortable light levels will vary with the individual. For example, the 60 year old worker requires 10 times more light to achieve good visibility than the 20 year old worker. Comfortable light levels will also vary with the task. The more rapid, repetitive and lengthy the task, the more important it is to have enough light. With these types of tasks, the eye is more vulnerable to fatigue and the worker to declined productivity.

Different colors of light will create different moods or atmospheres that will affect a worker's sense of well-being and level of productivity. Full spectrum fluorescent lights come closest to nature's light, imitating the color rendition of the noonday sun and adding a whole new sense of well-being to the office environment. This can be achieved by altering the lighting sources, or installing a special filter that can be placed between the lens and lamp of a fixture, or fit as a sleeve over each lamp. The many factors involved with deciding which lamps and/or lighting products and how to choose them will be discussed in detail.

Lighting for the workplace of today is distinctively different from what has been acceptable in the past. The computer display has replaced the telephone as the most important office tool in the 1990s. The average ambient light levels in most offices are too high, too inefficient and too costly. The trend now calls for reduced ambient lighting supplemented by adjustable task lighting. Recommended light levels for today's computerized workplace is 40-50 foot-candles for ambient light, as compared to 100 foot-candles or more in previous non-computerized offices. Many offices have no task lighting, yet task lighting systems are advanced, versatile and available to illuminate work surfaces and tasks without creating veiling reflections or glare on computer screens or work surfaces.

Many people inquire about the "best" colors for working on a computer display. The actual color of the letters and screen are a secondary consideration in this respect. More important is the contrast between the letters and the background. The combination that offers the maximum contrast is black letters on white background (like paper). This is very disappointing for many people, especially considering that they often have 16 million colors from which to choose! Be cautious of working on pale letters or very dark backgrounds in too bright of an environment.

Lighting a workplace for maximum efficiency is a nice concept. However, in the real world of budgets and bottom-lines, cost effectiveness is also a major consideration. The cost of energy, of new lighting fixtures, of retrofitting, of remodeling and more are all significant considerations which must be balanced to achieve the most for the money spent. Approximately 86 percent of the cost of lighting is energy consumption, while only 3 percent of the cost is the price of the lamp. Therefore, purchasing cheaper lamps does not necessarily indicate cost savings. A more prudent method is to purchase lamps that consume power more efficiently. These and other considerations were discussed in depth to allow for a thorough working knowledge of lighting within budget constraints.

Are the light fixtures equipped with standard prismatic lenses or grid-type lenses (parabolic louvers) that project the light out and down in the most efficient manner? Good fixture maintenance and energy efficient options are important. Achieving balance in creating effective task/ambient light levels for computer work is necessary. Checklists should be made to ensure that all lighting is ergonomically supportive of worker productivity before beginning work. Helpful reminders and current options were reviewed while focusing on the ultimate goal: to achieve worker-oriented lighting; lighting that will insure that the task can be comfortable and easily seen...and that the worker is working well.

Good workplace lighting involve three items: 1) Learning to observe the types of lighting available to the worker and to develop ongoing awareness of how this may or may not be working; 2) Identifying risk factors, such as glare and reflections, and the many options for correction of these factors; 3) Developing solutions that involve worker responsibility, administrative cooperation and caring, and realistic cost effective improvements.

GENERAL EYE CARE TIPS

Some of the practical recommendations regarding computer display viewing are:
- General eye care- the importance of regular and routine eye examinations, with emphasis on the computer using environment and working distances;

- Computer eyeglasses and frames- many optical companies are introducing lenses that are designed for the intermediate distance viewing situation. Your eye care professional will be able to guide you to the lens that is most effective in your particular viewing situation. There are also eyeglass frames which can be adjusted to provide an easier viewing area for computer work;
- Glare- there are many types of anti-glare screens for the monitor. The circular polarized glass generally tend to be the best and most expensive. The AOA has a listing available of those which have passed their qualification standard as being an effective screen;
- Working conditions- our eyes must adapt to our viewing environment. A poorly laid out workspace can lead to visual fatigue and eyestrain. Be sure that all items in the work area are easily seen and without excess glare or poor lighting;
- Breaks- taking visual breaks are easy to do and very effective in reducing eyestrain. A good rule of thumb is the "20/20/20" rule: every 20 minutes, take 20 seconds and look 20 feet away!

Eye health hazards have been touted for many years as a potential concern. As of today, there has been no proof that display use causes any type of eye health hazard. The electromagnetic radiation that comes from the computer monitor is well below all international standards and recommendations. There is more of a concern for this radiation that is emanated from the sides and rear of the monitor. Some eye care professionals feel that a UV protection is necessary for safety while working at a VDT. However, the research has failed to confirm that UV radiation has any effect on the VDT user. Most UV radiation drops off at about four inches from the front of the screen.

Contact lenses are also considered a concern for display users; however, with very little attention, those concerns can be easily addressed. Blinking is a problem for general VDT users and even more so if contact lenses are to be worn. Studies have shown that people blink less often while performing visually intensive tasks, and even less yet while viewing a VDT task. These results are probably a combination of concentration on the task and the position of the monitor. Most often monitors are higher in the visual field of view, therefore allowing the eyes to be open wider. This position is not our natural reading posture and will allow the eyes to dry out faster- with less blinking. This is even more critical for contact lens wearers. The use of lens re-wetting drops is recommended periodically during the day while using a VDT. It your doctor feels that wearing contact lenses is right for you, then that should be fine for VDT use.

The information presented here is very difficult to obtain for most employee populations. Eye doctors generally don't have the time, knowledge or inclination to discuss these issues while examining a patient in an office setting. Additionally, performing eye examinations in an office setting bears very little relationship to the working environment of today's office worker. A unique and popular software program, which actually performs a vision screening on the users own computer screen, is now available.

The effectiveness of the worker is dependent on adequate visual function, and visual function is dependent on appropriate lighting. The two areas are essentially inseparable in their interaction and critical in their effect on workplace performance. Knowledge in these fields is still growing and many professionals are poorly informed about this information which is essential for workplace effectiveness and productivity.

CHAPTER TWENTY ONE

Office Ergonomics Assessments Using the Internet

Ms Elizabeth Kestler, MS
Southwest Ergonomics Institute

Dr Henry Romero, CPE, ASP
Halliburton, Inc.

ABSTRACT

A method was developed for completing office ergonomic assessments using the Internet. The method is usable on a company Intranet as well as over the Internet by home office users. The method utilizes current understanding of ergonomics in offices and requires people to complete a series of forms including a picture of the workstation. Once these forms are submitted, a trained facilitator scores the forms and develops a set of recommendations for the individual. This software is specifically designed for non-complex situations, and particularly as a proactive measure. However, it is as effective reactively, if the musculoskeletal disorder symptoms are not too complex. The system was tested in a major energy services firm as well as across a group of people interested in musculoskeletal disorders in the office. The results demonstrate that the recommendations are essentially similar to those developed by an onsite assessment. In general, people were satisfied with the results. The benefits of the approach were also seen in time to complete the assessment being cut by 57% and the total cost reduced by 73%. However, it should be noted that people with complex musculoskeletal disorder symptoms or those with a history of dealing with musculoskeletal disorder symptoms, are generally dissatisfied as they prefer the human approach of an onsite assessment. In addition, an onsite assessment is capable of addressing other tasks performed in the office as well, for example, non-VDT work such as reading, handwriting, etc. In nearly 20% of the online assessments, an onsite assessment was recommended due to complex or severe musculoskeletal disorder symptoms. Even accounting for the additional cost of the onsite assessments, this method resulted in a cost reduction of approximately 50% and 37% reduction in time to complete the assessments.

1.1   INTRODUCTION

The recent issuance of OSHA 1910.900, Subpart Y, "Ergonomics" has renewed interest in ergonomics in all general industry, including office environments.

According to the United States Bureau of Labor Statistics repeated trauma injuries account for nearly 60% of all occupational injuries, with carpal tunnel leading the group. These disorders constitute the largest job-related injury and illness problem in the United States today. In 1997, employers reported a total of 626,000 lost workday musculoskeletal disorders (MSDs) to the BLS, and these disorders accounted for $1 of every $3 spent for workers' compensation in that year. Employers pay more than $15-$20 billion in workers' compensation costs for these disorders every year, and other expenses associated with MSDs may increase this total to $45-$54 billion a year. Workers with severe MSDs can face permanent disability that prevents them from returning to their jobs or handling simple, everyday tasks like combing their hair, picking up a baby, or pushing a shopping cart.

Southwest Ergonomics Institute developed software for the Internet that allows for the completion of an office ergonomics assessment using the services of a trained facilitator. This system is comparable to performing an assessment onsite with the services of a trained ergonomist. It differs from other software used to perform an assessment in that the results are not automatically scored and reported to the employee. This difference allows for two things: (1) an employer can control the results of the assessments and budget for improvements more easily and (2) current software products are incapable of evaluating pictures of workstation arrangements and accounting for the complexities of an office environment.

This paper describes the software and methodology for completing an assessment using it, and describes the result of a test recently completed. Additionally, this paper describes the costs and benefits of this method and compares it to the option that is most closely related to it, that of performing an onsite assessment.

In general, this method was tested in a major energy services firm as well as with a group of people particularly interested in office musculoskeletal disorders. The methodology consisted of having all of the people complete the online assessment forms including a satisfaction survey afterwards. The forms were scored, a report generated and submitted to the person, and then the person was requested to complete another satisfaction survey.

Some of the people completing the online assessment had an onsite assessment performed as well, in some cases by the ergonomist testing the online software, but also by other health and safety professionals. The results of the online assessment were compared to the results of the onsite assessment. In addition, those completing an onsite assessment were requested to complete a satisfaction survey that queried the person on the same elements as the online assessment satisfaction survey.

The results of the test showed the following. First, the online assessment process resulted in similar recommendations as the onsite assessment process. Second, the use of a workstation picture was not crucial to the validity of the assessment. Third, the online assessment process resulted in a 57% reduction in the time to complete the assessment and the total cost reduced by 73%. A complete cost-benefit analysis was developed for this software accounting for group and corporate rate breaks. Fourth, there are some people that are not going to be satisfied with this method, opting instead for the human touch of an onsite assessment. Fifth, certain people reported severe or a complex set of musculoskeletal disorders that, in the opinion of the ergonomist scoring the reports, required the completion of an onsite ergonomics assessment to verify the recommendations and ensure other factors did not need to be controlled. Sixth, there is some emerging research into computer workstation arrangements that could be considered.

1.2   PURPOSE FOR OFFICE ERGONOMICS ASSESSMENT

The office environment was long considered relatively safe from workplace injury. However, we have since learned that a significant number of back injuries and upper extremity musculoskeletal disorders are very common in office environments. Examples of upper extremity MSDs include: carpal tunnel syndrome, epicondylitis, cubital tunnel syndrome, De Quervain's disease, Guyon's canal syndrome, flexor tendinitis, pronator teres syndrome, and several others. In 1997, NIOSH commissioned a study completed by the National Research Council that demonstrates that there are several workplace factors that are associated with causing or exacerbating these MSDs. An ergonomics assessment is designed to evaluate these workplace factors and provide controls necessary to reduce their impact to the employees.

1.2.1   Basis for checklist items

The criteria for these checklists were taken from a wide variety of sources. Similar criteria can be found in the US Army ergonomics checklists, the ergonomics checklists, and checklists from several websites including the General Library task force and TIFAQ. There are several sources of checklists for office ergonomics assessments. Sources include the US army Maintenance website, the University of Texas Libraries website, checklists used by Halliburton, Inc., and the Job Evaluator Toolbox from ErgoWeb. Each of these sources list several important points that need to be evaluated in an office ergonomics assessment, and there is a great deal of common items. This set of common items, and the few different ones, was used to create a set of checklist criteria for people to respond to.

1.2.2   Impacts of OSHA 1910.900 Proposed

There is renewed interest in ergonomics since OSHA released 1910.900, Subpart Y, "Ergonomics", proposed, on November 23, 1999. This standard would apply to all general industry and would require analysis of all problem jobs upon the identification of a covered MSD. Many employers of large office workforces are concerned about how to analyze and implement controls in response to such a standard. In addition, small businesses are not exempt from the provisions of this standard, but often cannot afford a trained professional to assist them. OSHA understands that it will be difficult in office environments where the number of similar jobs could be large. OSHA has responded by saying that you can stop with one assessment if the recommendations look similar, but the burden of proof is on you as to when that is.

1.3   CURRENT ASSESSMENT METHODOLOGIES

There are several methods available and used for completing office ergonomic assessments. These methods include onsite evaluation, automatically scored and reported computer programs, and videotape or still photograph analysis. To understand how the method reported on in this paper differs from these, it is necessary to describe each of these methods and their associated advantages and disadvantages.

1.3.1   Onsite

The onsite assessment consists of a trained, qualified person using a set of checklists or other criteria to guide them through observations of the person using the computer and performing other tasks within the office. A good onsite assessment will examine the tools, furniture, writing utensils, software, monitor quality, ambient noise and lighting, workstation arrangements and layout, task lighting, and finally, the chair design and adjustments. This is the most comprehensive of the methods as it provides an opportunity to discuss concerns with the employee, experience their work environment firsthand, and account for other work activities beyond simply focusing on the computer. The disadvantage of this method is that it is very time consuming with the average, complete onsite assessment often taking an hour plus travel time to the person's workplace. Another disadvantage is that small businesses and home office workers often do not have the budget or resources to retain a trained, qualified person to perform the assessment.

1.3.2   Auto-scored Computer-based

The use of a computer system to perform office ergonomics assessments is not a new concept. Examples of computer-based systems include TOOL by Humantech, ErgoEASER by the Department of Energy, ErgoSmart evaluation module, and others. Each of these programs requires the user to enter data relative to their workplace and then it compares that information to an established model. Any deviations are pointed out to the user and an appropriate recommendation provided. In some instances, a hazard ranking is provided as well. The advantages of these programs is that a large number of people can perform an assessment in a relatively short period of time and at a reduced cost compared to performing an onsite assessment. There are several disadvantages of these programs. First, they do not account well for multiple systems needing changing. For example, if the seat height needs raising and the table height needs lowering, then the programs do not account for this well. In addition, some companies are not comfortable with a massive amount of employees being provided information that they need accessories, chairs, or other equipment, when the budgets may not be in place for such equipment.

1.4   VIDEOTAPE OR PHOTOGRAPH

Ergonomists have long used videotaping and still photography to perform assessments without travelling to the location. In the case of an office ergonomics assessment, a still photograph, or series of photographs can be used to show the workstation arrangements, equipment, accessories, ambient conditions, and other factors. The advantage of this method is that it does not require the ergonomist to travel to the location of the person to perform the assessment. The disadvantage is that a photograph often cannot depict the activities a person performs throughout their day, nor inquire about work methods.

1.5   OFFICE ERGONOMICS ASSESSMENTS ONLINE

Southwest Ergonomics Institute using a team of a doctoral-level, certified professional ergonomist with several years experience and a masters-level psychologist, also with several years experience, has developed an assessment methodology that combines some of the advantages of the onsite, computer-based, and still photograph assessments methods. The method uses the Internet to make it available to a wide variety of computer users, both corporate and personal users. The method specifically addresses the disadvantages of each of other assessment methodologies as well.

1.5.1   Purpose of the Online Assessment Process

The purpose of the Office Ergonomics Assessment Online process is to provide a vehicle for employers with large amounts of people working in offices to have a proactive assessment performed for all office employees. In addition, people working in small businesses or out of their homes do not have access to a trained, qualified professional to perform an ergonomics assessment. In each case, the software will collect the necessary data for a trained facilitator to perform an assessment and provide a series of targeted, personalized recommendations that will reduce the likelihood of developing a MSD. This can all be completed without hiring an army of trained professionals and spending significant amount of time and cost in travelling to the employee's locations. Furthermore, as more companies work towards telecommuting, this process will support the safety and health goals of those companies in a cost-effective manner.

The main purpose of this method is to perform proactive assessments in a timely, cost-effective manner. This process is not designed for people suffering from a complex set of severe MSD symptoms, though it provides a good method for identifying who these people are and alert trained professionals that an intervention is needed. The use of the software to identify who these people are ensures that they are identified in a structured manner that reduces the involvement of trained professionals until they are needed. The information is databasable and provides reports that can be used by management to budget for new equipment and accessories.

1.5.2   Methodology of the Assessment

Users are required to complete a series of forms designed to provide the facilitator with essential information regarding: (1) how people use the computer, (2) their discomforts while using the computer, (3) presence or absence of certain risk factors, (4) the presence or absence or arrangements of certain equipment, and (5) allows for uploading pictures of people using the computer workstation.

The forms are listed in Table 1 and a flowchart of the methodology is shown in Figure 1. The user begins by selecting the payment form and enters the pertinent data. After that, they can begin with any form and move to any other form though the software tries to guide them appropriately. The software does have validation built in to ensure no blank lines, etc. In addition, the user can leave at any time and can come back and begin with the first form that needs to be completed and submitted.

Each form asks the user a series of questions and provides them with options for responding. The options are specifically designed to be intuitive and descriptive, but not guide the user to a particular set of answers. The questions are specifically designed to investigate a wide range of physical layout issues, equipment and accessories, and the postures adopted while using the system.

The user is requested to submit a picture of them using their computer arrangement. This is one of the main features that separates this system from other computer-based evaluation software. The user is provided several options including uploading a digital photograph or snail mailing the picture. The user is cautioned to realize that their results will be delayed if they choose snail mailing.

The user is also requested to complete an evaluation of the assessment both prior to receiving results and afterwards. This feature is maintained beyond the testing period for purposes of continuous improvement.

Once the forms are received, they are printed for review purposes and the information is databased. A comparison table has been created that proceduralizes the scoring as much as possible. The procedure consists of comparing a list of recommendations to the responses provided by the user. Depending on the responses of the user, a particular recommendation is selected and included in the report. In addition, there are options for personalized, targeted recommendations to be added in at the discretion of the facilitator. The picture is then reviewed and compared against the recommendations. A Body Discomfort Survey is requested and if completed, the facilitator will use that information to gain insights into the impacts of the workstation to the employee. Also, if the facilitator sees significant or a set of complex MSD signs and symptoms, then an onsite assessment is recommended as is a medical evaluation. Finally, a series of workstation layout dimensions are provided based upon the user's reported anthropometrics. The report then includes a listing of sources for acquiring recommended equipment.

The entire process takes very little time to complete. On average, the test groups reported requiring between 30 and 60 minutes to complete and submit the forms. The facilitator needs approximately 10 minutes to review the responses and another ten minutes to score the results. Finally, another ten minutes is usually required to complete the report and get it sent back to the appropriate point-of-contact, often the user.

Table 1: Forms to be completed

Payment Form
Personal Computer Usage Information Form
Body Discomfort Survey Form
Risk Factor Analysis Form
Worksite Environment Survey Form - Part 1
Worksite Environment Survey Form - Part 2
Workstation Picture Upload Form
Assessment Evaluation Form
Post-Hoc Assessment Evaluation Form (After Receiving Results)

1.5.3 Limitations of this Process

Obviously, this process cannot resolve all office ergonomic situations as well as an onsite assessment. It was never designed to do so. This software assessment process is not expected to successfully provide an ergonomic intervention into people with complex, severe MSDs or physically complicated job tasks with the computer. However, it is anticipated that a proactive approach to ergonomics will limit the number of people with severe, complicated MSD signs and symptoms. In addition, people performing physically complex tasks with computers are relatively few and far between.

1.5.4 Advantages

The advantages of this software should be obvious. First, the time to complete the assessment is reduced tremendously when it can be completed within a total of one hour, including the employee's time, the facilitator's time, and reporting time. Second, there is no travel involved unless after reviewing the results, an onsite assessment is recommended. Third, large employers can survey all their employees in a relatively short period of time without hiring or mobilizing a large army of assessors. Fourth, small business owners and home office workers can take advantage of office ergonomics assessments without budgeting for an expensive consultant. Fifth, the information is electronically captured into a database allowing for searching and reporting findings to management.

Figure 1: Assessment Methodology

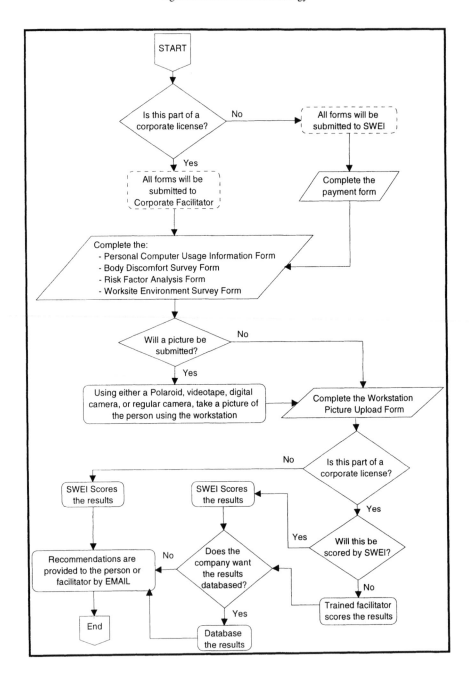

Figure 2: Risk Factor Analysis Form

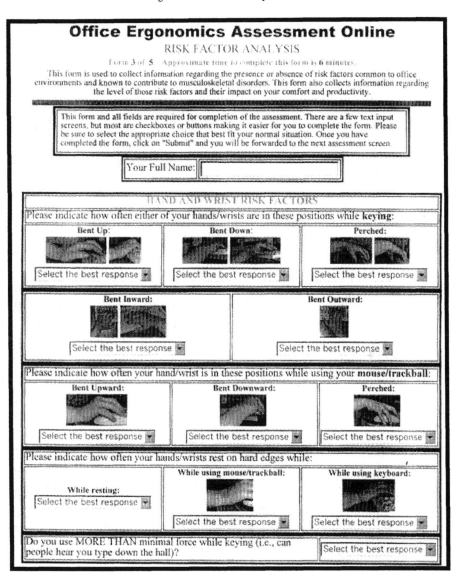

Figure 3: Risk Factor Analysis Form - Part 2

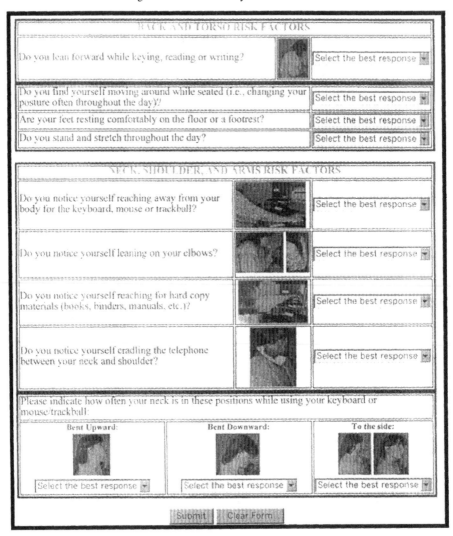

Figure 4: Worksite Environment Survey

Office Ergonomics Assessment Online

WORKSITE ENVIRONMENT SURVEY - PART 1

4 of 5. Approximate time to complete this form is 10 minutes.

The purpose of this form is to collect details concerning the arrangement, layout, and equipment in your office environment. This form is especially useful since the ergonomist is not physically visiting your workplace. This data will be used to determine the need for specific recommendations.

Also, this form is required for completion of the assessment. There are a few text input screens, but most are checkboxes or buttons making it easier for you to complete the form. Please be sure to select the appropriate choice(s) that best fit your situation. Once you have completed the form, click on "Submit" and you will be forwarded to the second half of this form.

Your Full Name: [   ]

Please select any workstation accessories you use regularly:

☐ Lumbar Support
☐ Padded keyboard wrist rest
☐ Padded mouse/trackball wrist rest
☐ Document Holder
☐ Task Lighting (lamp)
☐ Footrest
☐ Monitor risers (could be anything to support your monitor)
☐ Ergonomic keyboard
☐ Separate keyboard (for laptop users)
☐ Separate monitor (for laptop users)
☐ Other (please describe) [   ]

Question	Response
Is the seat height adjustable so you can place your feet firmly on the floor or footrest?	Select the best response ▼
Does the depth of the seat allow you to sit with your back firmly against the backrest without the front edge of the seat hitting your knees or feeling like you are going to fall out?	Select the best response ▼
Is the front edge of your chair seat rounded?	Select the best response ▼
Do you fully understand how to adjust your chair to fit you?	Select the best response ▼
Are the adjustments easy to reach and use while seated in the chair?	Select the best response ▼
Does your chair roll easily?	Select the best response ▼
Does the chair swivel?	Select the best response ▼

Figure 5: Worksite Environment Survey - Part 2

Does the chair have padded, adjustable arm rests?	Select the best response ▾
Does the chair have a five-star base?	Select the best response ▾
Is your chair comfortable?	Select the best response ▾
Does your backrest support your back?	Select the best response ▾

WORK SURFACE EVALUATION	
Is the work surface normally clean and organized?	Select the best response ▾
Is there normally enough space to get your work done?	Select the best response ▾
Is there a dedicated space for the computer?	Select the best response ▾
Is work surface height sufficient for you to get your legs under it?	Select the best response ▾
Is the work surface depth sufficient for you to sit close to your work?	Select the best response ▾
Is there adequate storage for your files, references, supplies, temporary work, personal items, etc.?	Select the best response ▾

COMPUTER WORKSTATION ACCESSORY EVALUATION	
Is a good document holder available and used?	Select the best response ▾
Is a padded keyboard wrist rest available?	Select the best response ▾
Is a mouse/trackball wrist rest available?	Select the best response ▾
Is an alternative keyboard (also called "ergonomic") design used?	Select the best response ▾
Is a glare screen provided?	Select the best response ▾
Is a comfortable footrest provided?	Select the best response ▾

Submit   Clear Form

1.6   COST-BENEFIT RATIO OF USING OFFICE ERGONOMICS ASSESSMENTS ONLINE

A cost benefit analysis was performed on this tool using several different scenarios. The reason for the variety of scenarios is that volume pricing is a component of the cost savings involved with the software. It is important to note that this cost-benefit analysis does not examine the benefits of having an assessment performed. The underlying assumption is that an assessment is to be performed, it is only a matter of the method. This cost-benefit analysis compares online assessments to onsite assessments and looks at savings in dollars and time, both important resources to most companies and individuals. The cost-benefit ratio for a single assessment is contained in Table 2.

Table 2: Cost-Benefit Ratio for a Single Online Assessment

Onsite Assessment	Assessment Time	Travel Time	Reporting Time	Review Time	Total Time	$/hour	Cost ($)
Facilitator	1	0.5	0.5	0	2	175	350
Employee	1	0	0	0.5	1.5	30	45
				Total	3.5		395
Online Assessment	Assessment Time	Travel Time	Reporting Time	Review Time	Total Time	$/hour	Cost ($)
Facilitator	0.25	0	0.25	0	0.5	NA	75
Employee	0.5	0	0	0.5	1	30	30
				Total	1.5		105

Cost Benefit Analysis	Dollars	Hours	
Onsite Cost	$395	3.5	
Online Cost	$105	1.5	
Savings	$290	2	
Benefit/Cost Ratio	276%	133%	{Savings/Online Cost}%
Cost Reduction %	73%	57%	{1-Online Cost/Onsite Cost}%
Payback Assessments	3.8	2.3	{Onsite Cost/Online Cost}

1.7   METHODOLOGY OF THIS TEST

The general methodology was to use two separate groups for testing the online assessment process. The first group consisted of office workers in a major energy services firm. The second group consisted of people subscribing to the Sorehand Listserve. The people in this group did not represent the most likely users of this software in that most are experiencing significant, severe MSD signs and symptoms.

The workers in the energy services firm were used to provide a corporate perspective to this test and also to provide a means for validating the results with onsite assessments. The Sorehand group is necessarily geographically dispersed as is possible with the Internet. In fact, subjects were from the United States, Canada, and South Africa.

The Sorehand and corporate group members were requested to complete the online assessments, including sending pictures of them using their computers. Once they were through with the assessment, they were asked to complete a post-assessment survey, prior to receiving results. Several of the corporate group had an onsite ergonomics performed to act as a validation for the online assessment.

Few people submitted a picture prior to their assessment being scored. However, several did submit pictures afterwards. This provided an opportunity to test whether the picture added significant value to the assessment.

Once the subjects received their reports, they were asked to complete a post-hoc survey in which they were asked questions regarding their satisfaction with the results.

1.8   RESULTS

1.8.1   Results of the Post-Assessment Survey

The users were queried as to their satisfaction and comprehension of the assessment process. The results of the survey are contained in Table 3.

1.8.2   Comparing online and onsite assessments

The results of the comparisons between the online assessments and the onsite assessments on the same person show the following:

- The onsite assessments had the same or less number of recommendations

- The type of physical or structural recommendations was the same. The online assessment incorporated personal activities, personal health information, and other activities into the recommendations, and the onsite assessment did not.

- The quality of the recommendations was improved in the onsite assessment. This was noted in that specific measurements and quick fixes could be implemented.

1.8.3   Comparing Picture versus no Picture

We began this study with the assumption that the results of these assessments are highly dependent on having a picture of the user at the computer workstation. However, very few of the subjects included a picture. There were a few that had submitted pictures, but their report was already scored. This offered the opportunity to update based upon observations from the pictures, thus demonstrating the value of the picture. In general, the following was found.

- The quality of the recommendations were slightly improved in that specific guidance regarding physical dimensions could be addressed when a picture was provided

- The number of recommendations increased slightly when a picture was submitted.

- The type of recommendations was improved as well, though not significantly in most cases.

Table 3: Post Assessment Survey Results

	Strongly Agree	Agree	No Opinion	Disagree	Strongly Disagree
This assessment was what I expected	4%	72%	12%	12%	0%
I understood the purpose of this assessment	28%	72%	0%	0%	0%
I consider this assessment an interruption	0%	16%	24%	28%	32%
I believe this assessment took too long	0%	8%	24%	48%	20%
I understood the information presented during this assessment	28%	60%	12%	0%	0%
I understood the questions asked during this assessment	28%	60%	8%	4%	0%
I believe this assessment focused on the things I wanted analyzed	8%	64%	16%	12%	0%
I believe there were additional things I needed assessed that were not covered	4%	32%	32%	24%	8%
I believe the program explained the assessment process adequately	12%	56%	16%	16%	0%
I expect the results of this assessment to make me more comfortable	8%	32%	56%	4%	0%
I expect the results of this assessment to make me more productive	12%	28%	44%	16%	0%
I understand the impact of ergonomics better as a result of this assessment	4%	36%	32%	28%	0%
I intend to recommend others participate in this assessment process	12%	36%	36%	12%	4%
This assessment was an enjoyable experience	8%	16%	56%	20%	0%

1.8.4 Satisfaction Survey Results

People were queried on their overall satisfaction with the assessment process prior to receiving results. The results of that survey are contained in Table 4.

Table 4: Satisfaction Survey Results

Post-Assessment	Excellent	Good	Fair	Poor
Satisfaction	22%	50%	28%	0%

1.8.5 Time to Complete the Process Survey

The users were queried as to how long the took to complete the online assessments. The results are shown in Table 5.

Table 5: Time to Complete Online Assessment

Time to	One hour	30min-60min	<15 min
Complete	29%	57%	14%

1.8.6 Results of the Post-Hoc Survey

The respondents were asked to complete a post-hoc assessment satisfaction survey after they received their reports. It should be noted that approximately 25% of the people involved in this study completed their post-hoc survey. The satisfaction survey results are included in Table 6 and Table 7.

Table 6: Post-Report Satisfaction Survey Results

Post-Report	Excellent	Good	Fair	Poor
Satisfaction	50%	0%	33%	17%

DISCUSSION

Although the sample size was smaller than desired, some important conclusions seem evident. Namely, immediately after finishing the assessment, the users responded with the following:

- The online assessment process was what they expected (76%)

- They understood the purpose of the assessment(100%)

- They did not consider it a disruption (60%)

- They felt the process did not take too long (68%)

- They understood the questions asked (88%)

- They felt the assessment focused on issues they wanted analyzed (72%)

- They felt the process was explained adequately (68%)

- They had no immediate opinion on the process would make them more comfortable (56 no opinion)

- They had no immediate opinion on whether the process would make them more productive (44% non opinion, even distribution)

- They had no immediate opinion on whether the process increased their understanding of ergonomics (even distribution)'

- Few claimed the assessment was an enjoyable experience (24%)

- The average time to complete the assessment was between 30 minutes and 60 minutes, with most reporting closer to 30 minutes

- Nearly everyone was satisfied with the process prior to the results being disseminated with the lowest score being fair on satisfaction (28%)

- The online assessments did not vary much from the onsite assessment, though perhaps the online assessments were able to recommend personal factor controls easier since the assessor was remote from the employee

- The online assessments were completed faster than the onsite assessments

- A picture does the improve the results slightly, though not as much as anticipated

- When combined with the Body Discomfort Survey, it acts as a good screening tool for identifying those requiring an onsite assessment

- People were split on whether the results were better than expected (34%)

- People felt the recommendations were appropriate (67%)

- People felt the results were provided with enough detail to be implemented (67%)

- People were split as to whether they had the resources to implement the recommendations (43% did)

Table 7: Post-Hoc Assessment Survey Results

	Strongly Agree	Agree	No Opinion	Disagree	Strongly Disagree
Results were better than expected	17%	17%	33%	33%	0%
Results were of sufficient detail to implementation	50%	33%	17%	0%	0%
I intend to implement most or all of these recommendations	50%	17%	0%	33%	0%
I have sufficient resources to implement most or all of these recommendations	14%	29%	14%	43%	0%
I believe these recommendations are appropriate for my situation	50%	17%	0%	33%	0%
Report was better than expected	33%	17%	33%	17%	0%
Resources provided on Office Ergonomics Assessments Online were useful	33%	17%	50%	0%	0%
The pictures and the Computer Workstation Dimensions were useful	17%	33%	33%	17%	0%
The assessment was easy to read and understand	33%	50%	17%	0%	0%
I felt that the assessment was based upon the latest understanding of ergonomics	33%	33%	17%	17%	0%
I intend to recommend others participate in this assessment process	33%	17%	50%	0%	0%
The links in the report were useful	17%	0%	83%	0%	0%
Report personally developed for me	33%	33%	33%	0%	0%
I believe there were things I needed assessed that were not covered	0%	33%	33%	33%	0%
I expect the results of this assessment to make me more comfortable	17%	50%	0%	33%	0%
I expect the results of this assessment to make me more productive	17%	17%	50%	17%	0%
I understand the impact of ergonomics better as a result of this assessment	17%	33%	33%	17%	0%

- Most people thought the report was what they expected or was better than what they expected (83%)

- Everyone either felt the additional information was useful, or had no opinion

- Only 17% of the respondents felt the provided workstation dimensions were not useful

- Nearly everyone agreed that the report was easy to read and understand (83%)

- Only 17% of the respondents felt the assessment did not reflect the latest concepts in ergonomics

- Half of the respondents said they recommend the process to others, with the remainder not expressing an opinion

- People generally had no opinion on whether the links to suppliers contained in the report were useful 83%)

- People generally felt the report was personally developed for their situation (66%)

- Two-thirds expected the results to make them more comfortable and only 17% thought the results would not make them more productive.

In general, the results for the online assessments were similar to those of the onsite assessments. The main advantage to onsite assessments is that they provide an opportunity for one-on-one training and for some quick fixes. Other than that, the results and perceptions are remarkably similar. It should be noted that onsite assessments require substantially less labor and time. We did note that pictures were not as crucial as originally anticipated. Finally, the online assessment acted as good screening tool for identifying people that needed an onsite assessment. The use of this tool would help to manage the costs of onsite assessments while directing this scarce resource more appropriately.

1.9   CONCLUSIONS

This study demonstrates that this online ergonomics assessment process is nearly as effective as an onsite assessment process, for the majority of the population. In addition, it allows for employers with a large office workforce to meet their goals of assessing each individual workstation in a reasonable amount of time and labor cost. Also, small businesses and home office/computer users can benefit as well where previously they did not have the resources to secure the services of a trained ergonomics professional. This process works well as a screening tool ensuring that those with significant MSD signs and symptoms are identified and appropriately resolved. Logically, a picture of the workplace would seem to make the assessment more personalized and accurate. However, this study did not support that conclusion. It should be noted that this process allows for databasing of the data that puts more of this pertinent data in the hands of the management and other decision makers. Finally, while this software provides for proactive analysis of workstations that are nearly as effective as onsite assessments, it does so at 57% of the cost and two-thirds the time.

1.10 REFERENCES

Below is a list of Internet and traditional references that contributed to the development of the Office Ergonomic Assessment Online tool and the chapter.

Internet

Donkin, R. (1994) *An Argument for Ergonomic Workstations: Increased Productivity.*Reprint. http://www.tifaq.org/articles/ergo_increased _productivity-apr94-richard_donkin.html

Johnson, P. (1994) *UCS/UCB Ergonomics Program: Pointing Device Summary* http://www.me.berkeley.edu/ergo/tips/pdtips.html

Joyce, M., Marcotte, A., Calvez, V., Barker, R., Crawford, P., Klinenberg, E., Cogbum, C., and Goddard, D. (1997) *Preventing Work-Related Musculoskeletal Illness Through Ergonomics: The Air Force PREMIER Program Volume 3A: Level I Ergonomics Methodology Guide for Administrative Work Areas.* Occupational and Environmental Health Directorate. http://sg-www.satx.disa.mil/hscoemo/Level 1/select1.htm

Poynton, C. (1997) *Reducing Eyestrain from Video and Computer Monitors.* http://www.infoamp.net/~poynton/notes/ reducing_eyestrain/

Sheehan, M. (1990) *Avoiding Carpal Tunnel Syndrome: A Guide for Computer Keyboard Users.* Reprint. http://www.indiana.edu/ ~ucsstaff/ cts.html.

Computers and Eye Strain. American Academy of Ophthalmology. http://www.eyenet.org/public,faqs/computers_faq.html

Ergonomics 101: The Evolution of Sitting . http://www.proergo.com/ ergo101sit.htm.

Office Ergonomic Assessments Online. Southwest Ergonomics Institute. http://www.ergodoc.com

Office Ergonomics Practical Solutions for a Safer Workplace. Washington State Department of Labor and Industries, http://www.lni.wa.gov/ wisha/ergo/officerg/toc.htm

Fitting the Job to the Worker: an ergonomics program guideline, Washington State Department of Labor and Industries, http://www.lni.wa.gov/wisha/ergo/veg/vegtoc.htm

Workstation Ergonomics, http://www.lib.utexas.edu/Pubs/etf/exhibit.html

Literature

Kasdan, M., ed. .*Occupational Medicine Occupational Hand Injuries.* Philidelphia: Hanley & Belfus, Inc., 1989.

Mill, W. *RSI Repetitive Strain Injury.* San Francisco: Thorsons,1994.

Moore, S., Garg, ed. *Occupational Medicine Ergonomics: Low-Back Pain, Carpal Tunnel Syndrome, and Upper Extremity Disorders in the Workplace.* Philidelphia: Hanley & Belfus, Inc., 1992.

Pascarelli, E., Quilter, D. *Repetitive Strain Injury: A Computer User's Guide.* New York: John Wiley & Sons, Inc, 1994.

Ostrom, L., Gilbert, B., & Wilhelmsen, C., (1991) *Summary of the Ergonomics Assessments of the Selected EG&G Idaho Work Places,* EGG-2652

Ostrom, L. (1993) *Creating the Ergonomically Sounds Workplace,* Josey-Bass, Inc., San Francisco, CA.

Crouch, T., (1995) *Carpal Tunnel Syndrome & Repetitive Stress Injuries: The Comprehensive Guide to Prevention, Treatment & Recovery,* North Atlantic Books, Berkley, CA.

US Dept of Health and Human Services, (1995) *Cumulative Trauma Disorders in the Workplace: A Bibliography,* DHHS(NIOSH) Publication No. 95-119.

CHAPTER TWENTY TWO

"Ergo Nightmare" ... and Beyond

Joy L. Sisler, CPE
American Ref-Fuel Company

In March of 2000 I gave my presentation at the 3rd Annual Applied Ergonomics Conference in Los Angeles, California. The audience response was overwhelming! There have been many inquiries about the program and the videotape - some inquirers were Boeing, Honda, Ford, NASA and Gillette. It has been a pleasure sharing information with fellow HSE folks and I look forward to the continued sharing with each other in the future.

We updated our program by expanding the education to our plants, with a hands-on approach. Some of these experiences are included in the following updated version of my prior presentation.

Introduction

American Ref-Fuel Company builds, owns and operates waste-to-energy plants. We are very proud that four of our six plants have received OSHA's Voluntary Protection Program Star classification for workplace safety. This elite status has been attained by fewer than one in 10,000 of the worksites monitored by OSHA. We feel that having reached this lofty goal four times is truly an honor and a tribute to our employees' total commitment to safety. It proves that we live our philosophy statement, which says, in part, *"Total Safety* is the ongoing integration of safety into all activities with the objective of eliminating injuries and improving performance... Nothing is more important than safety."

With such a strong safety culture, we found it troubling in 1992 and 1993 when three Ref-Fuel employees in the company's corporate headquarters in Houston, Texas, developed carpal tunnel syndrome. Each required corrective surgery. In 1994, American Ref-Fuel's Health, Safety, and Environmental ("HSE") committee was formed, including our Vice President, Safety and Environment, Andrew Szurgot, as the executive sponsor. He was joined by at least one member from each major department. In order to help prevent more injuries, the company decided to investigate office ergonomics. The first step was for the HSE committee to incorporate into its Mission Statement the following primary objective: "...Consistent with Ref-Fuel's Total Safety Philosophy to provide a safe and healthy workplace, devise and maintain Ergonomic and Safety Programs in the Houston office."

Reporting Work-related Injuries

Mr. Szurgot issued a memo reminding our corporate employees to immediately report work-related injuries to their supervisors. All workplace hazards and even near misses were also to be reported. Our goal is to continually raise the awareness about workplace injuries and safety hazards, encouraging all employees to confidently step forward in order to take appropriate precautions.

Development and Implementation

We incorporated office safety programs into our collaborative culture. Our insurance carrier offered training programs for employees to become Certified Ergonomists. Two HSE committee members, (one who had developed carpal tunnel injuries) and I, stepped forward to get the certification training.

As new Certified Ergonomists, we were ready to make a difference! We weren't entirely sure where to start, so we began by targeting the high-risk (i.e., clerical) users. Next, we issued Workstation Evaluation Surveys. Upon receiving feedback from the surveys, we evaluated each employee's workstation. We determined the following plan of action:
1. There was a need to install keyboard trays that were adjustable and wide enough for mousing.
2. There was a need to provide wrist rests.
3. Phone handset cradles and headsets were needed.
4. Ergonomic chairs were needed throughout the office.

Management Support

In order to convince management to provide funds for updating the ergonomic safety of workstations, we drafted a plan with a budget that included projected savings on medical expenses and employee lost time. We compiled a list of equipment needs with prices and allocated costs to the various departments. It was not difficult to convince management to invest in the equipment to keep employees ergonomically safe, especially when they learned from our insurance carrier, that a typical expense for one repetitive injury can cost employers anywhere from $20,000 to as much as $150,000. It also helped to have a company executive acting as the executive sponsor on the HSE committee. Once we received management's approval, we researched different vendors and purchased the required equipment.

Ergonomic Equipment Installation

Most employees welcomed the equipment installation. They appreciated the fact that their employer was concerned enough to provide the proper equipment. A few employees, however, were not as keen on the changes required – they liked their workstations in existing configurations. In order to convince them, we had to reiterate that it's the cumulative repetition that could eventually cause them pain and disability. We asked those who were still reluctant to at least try changing for awhile. Most agreed to a temporary arrangement. Eventually, they began to realize that proper ergonomics is the right thing to do.

Training Employees

After installation, we educated the employees on why and how to use the new equipment. We used video presentations to start our employee training. We viewed several different videos available on office ergonomics and purchased the one we agreed was best for our needs. Twice each year we showed this video to all corporate office employees. This training provided:

1.   Understanding of what ergonomics is about
2.   Awareness of specific types of ergonomic needs
3.   Prevention of future cumulative trauma disorders (CTDs), repetitive stress injuries (RSIs) and musculoskeletal disorders, (MSDs)

During the training, we explained about CTDs, RSIs and MSDs and we described carpal tunnel syndrome and other typical conditions.. We presented photos illustrating ergonomic problems and potential corrections with captions such as, "What's Wrong With This Picture?" and "What's Right With This Picture?" Using our own employees as models enhanced interest in this training tool.

Involving the Information Systems Department

After installing the new equipment and initiating employee training, we asked the Information Systems (IS) department to become more involved in the ergonomics of workstation setups. We trained the "Help Desk" person and two other IS personnel. This training was more in-depth than the typical training for employees. Our goal was to educate them so they could correctly set up new ergonomic workstations. In addition, when they were called to help adjust existing workstations, they needed to understand the rational behind the suggested changes.

Reconfiguration of Workstations for Laptop Users

American Ref-Fuel has many laptop computer users. When using laptops at their desk, however, many of these people were not very comfortable. We could see that modifications were needed. For some users, we installed a keyboard tray, full-sized keyboard and mouse, and a full-sized monitor – all connected to their laptops. These modifications improved their posture as well as maintaining a neutral keyboard position and proper mouse position. In one instance, the user did not want the full-sized monitor, so we used risers to get the laptop monitor to the appropriate eye level.

Our laptop users are now ergonomically happy campers!

Different Mice

Not all mice are created equal. Mousing has become a major part of computer applications. Therefore, CTDs have started to occur in the fingers, wrists, elbows and shoulders of mouse users. Moving the mouse to the same level as the keyboard was critical but still not the entire answer for some. User Number 1 was having trouble with elbow pain. Switching this person to a large marble rollerball mouse eliminated the pain. User Number 2 had an arthritic condition in the thumb and palm. We tried many different mice to determine proper fit and comfort factors. Finally, a mouse was identified that not only fit the user's hand, but also helped increased comfort immensely. Interchanging mousing hands can also been helpful. Although most people are not exactly ambidextrous, mousing with either hand is a skill that can be acquired with a little practice and patience.

Taking Breaks

We emphasized to our employees that today's technology makes it easy to just keep plugging away on the keyboard and mouse without breaks. However, we explained why it is imperative to take breaks about every twenty minutes in order to give the muscles a break and help the circulation.

We further explained the importance of the following stretching exercises to maintain healthy muscle tone and joints, and improve circulation:

For the wrists and hands:
- Spread fingers and extend for a good stretch
- Stretch wrists by gently pulling fingers back
- Repeat at least twice

Body stretches:
- Stand up and reach for the ceiling
- Take a deep breath and relax
- Repeat at least twice

The Eyes Have It

We also reminded the employees that while it's easy for some people to sit and stare at the computer monitor for hours, such habits can take its toll on the eyes. Without breaks, eyestrain and headaches can result. It's important to take breaks about every 20 minutes, focusing outward by at least 20 feet. Eye exercises also help – as if looking at a large clock, moving the eyes in a clockwise direction and then counterclockwise can be helpful. Frequent blinking can help relieve the eyes, as can eye drops for restoring moisture.

Bifocal and Trifocal Challenge

Many of the computer users in our company wear bifocal or trifocal eyeglasses. This poses another ergonomics challenge at the workstation. Lowering the monitor usually helps, but it depends on which vision segment they are looking through to see the monitor. The number of hours per day the user spends in front of the monitor can also be a determining factor in the solution. For example, if a person with bifocals or trifocals spends many hours a day in front of the monitor and is having problems focusing, one recommendation would be to adjust the workstation. However, if adjusting the workstation does not remedy the problem, then task glasses might be in order. Task glasses are specifically made for the distance to the computer screen. Wearing task glasses can eliminate the user's problems of trying to figure out which lens to focus through, especially when using a large monitor.

We have had several employees get task glasses and they are getting along very well.

Pain In the Neck

We noticed that many employees cradle the phone between their ear and shoulder. Doing this consistently for long periods of time can cause a real pain in the neck. This pain can radiate into the shoulder. It is not difficult to figure this one out – our necks were not designed to hold our heads at that angle for any length of time. A person's head weighs about the equivalent of a 10-pound bowling ball. Tilting this weight at an angle puts a lot of stress on the neck muscles and vertebrae.

For those who habitually cradled their phones, we recommended:
1. Use a speaker phone
2. Purchase a handset cradle
3. Purchase a headset

We have found that frequent phone users do best by either using the speakerphone or purchasing a headset. Good headsets are priced starting from about $79 and it sure beats that pain in the neck!

Exercises for the Neck and Shoulders

Here are the neck and shoulder exercises we recommend to our company's employees. These exercises can be done either sitting at a desk or standing up:

1. First, relax and drop the shoulders and take a deep breath.
2. Raise shoulders toward ears, then release and relax the shoulders.
3. Tilt head forward far enough to feel the stretch in the neck and between the shoulder blades.
4. Tilt head from side to side, far enough to feel the stretch in the neck.
5. Turn head to the left and then to the right.
6. Don't forget to breathe throughout this routine.

Follow Up with Employees

The final and continuing step of our office ergonomics program is responding to employee questions and concerns. We frequently circulate e-mails and memos to remind employees that if they have any issues, concerns, pain, or discomfort of any kind, they can call on their in-house ergonomists to discuss the issues.

When an employee does have a complaint, the first thing we do is an evaluation of the workstation and, if necessary, get a history. We ask whether this is the first time they have experienced this pain? How often does it occur? When does it occur? Have they ever been in an accident or had an injury in the area that is bothering them? Do they have a condition that might be related to this problem, such as arthritis? After doing an in-depth evaluation, we determine whether it is an ergonomics issue related to their workstation. If so, immediate action is taken to correct the situation. The person's hobbies and home projects are also important factors to consider. For example, a tennis player can have hand, wrist, elbow, and back and knee problems. We examine the types of projects they are involved in at home (especially at those involving repetitive motion). If applicable, we have them describe their home workstation.

Why The "Ergo Nightmare" Video?

After two years of using the same ergonomics training video, we believed a fresh approach would be more effective with our employees. We viewed many different commercial videos, none of which were very memorable. The HSE committee decided to try our hand at making our own ergonomics video. We established a subcommittee for this endeavor and the fun began! In order for us to be able to get our "day jobs" done, we agreed to meet at 6 a.m. each day that we worked on the project. Donuts and coffee were provided to entice our employee "actors" out at this early hour. Our first two sessions were mainly brainstorming – determining what we wanted to accomplish. Our three-fold goal became to educate, entertain and make it memorable. Our theme was "Ergo Nightmare", which opened with a takeoff on "The Blair Witch Project." The videotaping took several sessions to capture all of the scenes we wanted. For the most part, scripts were not scripted; we just used the ideas and played it by ear. Once we had plenty of video footage, it was time for editing. Our editing team cut, spliced and added music.

Then it was "Show Time!" We started promoting the video with posters and e-mails to get everyone's attention. We unveiled our production on Halloween morning to a packed house in our company's largest conference room! We played spooky music in the room to get everyone intrigued. And of course, donuts were once again provided. Lights out! And, action! Our audience loved the video! We accomplished all that we set out to do -- educate, entertain and make it memorable.

The Ergo Nightmare Video

The nightmare starts out with our group running in the dark behind our building. Then we are going through a jungle in the office. A visitor goes to an office to discover an ergo nightmare – a person with bugging eyes from looking at a monitor for too long. Next, a visitor goes to an office to discover someone cannot move their neck, from cradling the phone. Then, another employee is visited and is exhausted with too much work to do without any breaks. And finally, an employee leans back in his chair and then can't stand up because his back hurts.

In the video, these nightmares were 'dreamed' by our Vice President of Safety and Environmental. He wakes up, drives into the office to report that he has had a nightmare about the ergonomics program falling completely apart.

In my video role as the "Ergo Fairy Godmother" - with my magical chimes I visit the individuals who had appeared in his ergonomics nightmare and helped them to understand proper ergonomics at their workstations.

The video ends on a serious note, explaining that the light-hearted approach was only used as an attention-grabbing device to reinforce the importance of proper ergonomics, taking breaks, and avoiding ergo nightmares.

Ergo Fairy In Training

In order to make it interesting at the plants, we developed the "Ergo Fairy In Training" program. First, we identified the safety supervisors of the plants and provided them with the same training as office workers. Second, we did ergonomic evaluations and recommendations, accompanied by the supervisors, thus, the nickname "Ergo Fairy In Training". This hands-on technique provided an in-depth understanding for the supervisors. Once the report was written and submitted to the safety supervisors, it was pretty clear as to what action items needed to be taken care of. Although the "Ergo Fairy In Training" seems light-hearted, it gets the point across and encourages a little camaraderie.

Conclusion

Since the implementation of our office ergonomics program, we have had no work-related injuries in our office. We are continuing to expand our program to include workers in the plants.

Change is inevitable. At American Ref-Fuel, we strive to build enthusiasm so all employees want to cheerfully participate in our ergonomics program, as well as in our total safety philosophy.

CHAPTER TWENTY THREE

Questioning Office Ergonomic Guidelines

Dennis R. Ankrum, CIE

You've got a problem with office ergonomics. Or maybe you just want to prevent problems. You've attended a couple of ergonomics seminars, but you're not an ergonomist. What do you do? Many people turn to guidelines. But which ones?

That guideline you're considering might be based on outdated science, or it might not even apply to your jobs or people. Before you invest your company's money in new equipment or employee training, you need to ask some questions about those guidelines you're preparing to use.

This chapter will ask some of those questions and provide some examples of how guidelines can be misapplied.

1. DO THE AUTHORS KNOW WHAT THEY ARE TALKING ABOUT? DO THEY HAVE ANY HIDDEN AGENDAS?

Guideline makers run the gamut from individual ergonomists to government committees to companies selling ergonomic products. The quality of the guideline depends not only on the capabilities of the makers, but also on their goals.

Many, but not all, mass-produced, for-profit guideline publishers have to keep their customers in mind. Most often, the customer is the employer. Some publishers may fear that if they recommend ergonomic fixes that are too expensive, the employers won't buy their guidelines. Be careful if a guideline concentrates too much on changing the worker to fit the job, and not vice versa.

Local, state, national and international standards are usually developed by committees composed of many viewpoints: labor, employers, government, ergonomists, and suppliers of ergonomic products. Because of this, committee-built guidelines tend to represent the minimum recommendations that everyone can agree on, rather than the optimal level of practice. Check the introduction of the standard, where the minimum-optimum tradeoff may be elucidated.

Manufacturers of ergonomic products and trade and employee organizations often publish guidelines. They do this for many reasons, including limiting product liability, promoting products, public relations, and sometimes good ergonomics. Some of these guidelines are ghost-written by ergonomists, but others are put together by marketing or legal departments. If not done by ergonomics professionals, their recommendations are too often cobbled together from questionable published sources that aren't scrutinized and put into the right context.

There also may be a tendency to emphasize solutions involving their own products. If you use guidelines from these sources, check the credentials of the authors, if you can.

Ergonomists are often involved in guideline writing on behalf of their own agendas, which may be the same as the ones described above. However, if the guideline is based on one actually used by the ergonomist in their own work, the results can be as good as the ergonomist is.

2. DOES IT EXPLAIN THE "WHYS" FOR THE RECOMMENDATIONS?

Try to figure out exactly what the recommendation is trying to accomplish. A guideline should explain the reasons behind its recommendations well enough so that you know when you should ignore or change them. A good rule is to never follow a guideline if you can't explain the principles behind it.

"Keep forearms parallel to the floor."

The purpose for this recommendation is to ensure straight wrists. But keeping the forearms parallel to the floor results in straight wrists only if the keyboard is horizontal and at the same height as the elbows. Most guidelines don't include this qualifier. If keyboards are lower or higher than this (and there are some good reasons for lower keyboard heights, such as negative tilt keyboard trays), then the statement gives bad advice because it causes bent wrists.

"Keep the feet flat on the floor or on a foot rest."

Although some people think that the feet should be flat on the floor at all times, the reason behind the recommendation is to avoid excessive pressure on the underside of the thighs, and to ensure that the user can use the floor for support when shifting postures. Because thigh pressure risks don't outweigh the ergonomic value of shifting posture frequently (a.k.a. fidgeting), the guideline should read "CAN the feet be placed flat on the floor or on a foot rest?"

3. DO THE TASKS HAVE ANYTHING TO DO WITH WHAT YOUR WORKERS DO?

Most office ergonomic guidelines don't tell you to what kind of jobs or tasks they apply, but if you look closely, it's often continuous data entry. They might work for your data entry jobs, but may not make sense for anyone else. For example, a fully articulating, height adjustable keyboard with all the bells and whistles will only get in the way of someone who uses their computer only to read bits of e-mail. Moreover, the high-risk task of handwriting is completely overlooked in computer-focused guidelines.

Different jobs truly require different solutions. Before instituting a change, decide what you want to accomplish. Sometimes what is appropriate for one task will make things worse for another.

4. DO THE DIMENSIONS IT RECOMMENDS FIT YOUR PEOPLE?

Unfortunately, most guidelines assume an average-sized group of workers.

"The seat depth should be 17 inches."

Here's another instance where understanding the reasoning helps. An adequate seat depth both supports the thighs and allows the user to sit back far enough to use the backrest without creating pressure on the backs of the knees.

Here's a slightly extreme, but true, example. An Asian university needed to purchase chairs. They looked at several national standards and found almost universal agreement for a seat pan depth of 17 inches, which just happened to be the specification for most of the chair choices that were available. After the chairs were delivered, they discovered that almost none of their students could sit with their lower backs against the backrests.

Workstation dimension recommendations are often based on anthropometry, the study of body measurements. The custom is to design products to fit from the 5th percentile female dimensions to the 95th percentile male dimensions.

Many people assume that there exists a "fifth percentile female" and a correspondingly large male. However, such a person is unbelievably rare. For all dimensions on one person to be proportionately small (or large) involves a probability much less than 5 out of 100. That is because the percentiles refer to individual body segments, not standing height. If you look around, you will notice that people with long shins tend to have shorter thighs, and vice versa. Things even out. Therefore, while 5 out of 100 users will be too "long" or "short" for a "typical" seat depth, others will be all wrong for "typical" armrest heights, but fine for everything else.

Chris Grant, a Michigan office ergonomist, does an exercise in her workshops that gives an indication of how many people fall outside the 5th through 95th percentile range in areas that affect computer workstations. She has attendees measure four body segments that are relevant to seating fit, plus standing height. About 50% of the participants fall outside the 5th percentile female - 95th percentile male range for at least one body dimension.

The implications here are important. To make sure more than about half (according to the Grant results) of your workers fit your chairs, you MUST have some workstations that are smaller, and some that are larger, than the fabled 5th-95th percentile. This means workstations that exceed ANSI (Human Factors Society, 1988) recommended adjustment ranges.

5. IS THE GUIDELINE BASED ON SCIENCE, OR JUST PREFERENCE?

Some guideline recommendations are based on preference, and some on science having to do with wear and tear, anthropometrics, or carefully measured discomfort. Preference data are often crude and can be influenced by equipment, habit and environmental factors. Sometimes preferences get confused with science.

One example is a study published in 1982, "Preferred VDT workstation settings, body posture and physical impairments," by Grandjean (Grandjean *et al.*, 1982). The researchers installed new, adjustable workstations and measured how they were adjusted. One finding was that the preferred viewing angle from the eyes to the center of the screen averaged 9 degrees below horizontal eye level.

In light of recent research that shows the benefits of lower monitor positions (see Ankrum and Nemeth, 1995 for a summary), the preference doesn't make sense. That is, until you look at a later publication of the same data, "VDT Workstation Design: Preferred Settings and Their Effect," (Grandjean *et al.*, 1983). With the new workstations, the complaints of strong, annoying screen reflections was reduced from 32% to 12%. The lighting was not optimised for VDT work and lower monitors would have resulted in even more glare on the screen.

In addition, to keep the screen at a right angle to the line of sight requires tipping it back one degree for every degree the center is below eye level. Because the monitors in the study could only tip back 10 degrees, a line-of-sight lower than 10 degrees to the center of the screen would have resulted in a condition with the top of the screen closer to the eyes than the bottom. That's uncomfortable and can increase eyestrain and postural discomfort (Ankrum *et al.* 1995).

Grandjean's recommended workstation settings have been cited repeatedly without an explanation of the limitations of the study.

A positive aspect of Grandjean's study was that it was done in a real life work environment, with real workers. Too often preferences are gathered from college students who have no experience in the area being surveyed.

6. ARE THE RECOMMENDATIONS OVERLY PRECISE?

"Keyboard heights should be adjustable from 27.7 to 33.5 inches."

There are no cliffs in ergonomics. One inch farther will not cause a fall to the death.

Translating from centimeters to inches can get ridiculous. For example, a 70-85 cm recommendation can be converted to 27.66-33.5 inches. That implies much more precision than the science supports.

The less familiar we are with the principles, the more precise we tend to be in following directions. Any time you see decimal places after a number, be suspicious.

7. DO YOUR WORKERS REBEL WHEN YOU TRY TO IMPLEMENT THE GUIDELINES?

"Maintain the ergonomically correct posture. See Figure 2."

If you have to fight to get workers to follow the guidelines, then there might be something wrong with them. The guidelines, that is, not the workers.

Some guidelines include diagrams that dictate "proper" posture. "Figure 2" then shows a worker with every body angle at precisely 90 degrees. One guideline even insists that the toes must face forward.

Workers naturally rebel. They only assume the "ergonomically correct" posture when the posture police are in the area.

Many guidelines talk about a single posture as if workers shouldn't move around, but the preponderance of ergonomic evidence supports plenty of movement ... postural variety. The variations in posture should generally be within a range, a certain distance from "neutral," but even awkward postures can be desirable if they are voluntarily assumed and not sustained.

When workers balk at following suggestions, ask them why. Their reasons might make sense. The best behavioral changes are usually achieved by explaining the reasons for the guidelines and being flexible.

8. DOES IT WORK?

The ultimate criterion for judging a work environment is not how well it conforms to the guidelines. The bottom line is how well it facilitates the ability of the person to perform their work, effectively and without injury.

REFERENCES

Ankrum, D.R., Hansen, E.E. and Nemeth, K.J., 1995, The Vertical Horopter and Viewing Distance at Computer Workstations, *Symbiosis of Human and Artifacts*, eds. Anzai, K., Ogawa, and Mori, H., Amsterdam: Elsevier.

Ankrum, D.R. and Nemeth, K.J. 1995, Posture, Comfort, and Monitor Placement. *Ergonomics in Design*, 3, 2, pp. 7-9.

Grandjean, E, Hunting, W. and Nishiyama, K. 1982, Preferred VDT workstation settings, body posture and physical impairments. *Journal of Human Ergology*, 11, pp. 45-53.

Grandjean, E, Hunting, W. and Pidermann, M. 1983, VDT Workstation Design: Preferred Settings and Their Effects. *Human Factors*, 15, 2, pp. 161-175.

Human Factors Society, American National Standards Institute, 1988, A*merican National Standard for Human factors Engineering of Visual Display Terminal Workstations*, Santa Monica, CA: Human Factors and Ergonomics Society.

CHAPTER TWENTY FOUR

Resolving RMI's in Manual Insurance Policy Assembly

James A. Boretti, CSP

1   INTRODUCTION

This case study stems from a continuing problem faced by major insurance companies. At issue here is the number of repetitive motion injuries (RMI's) reported in the insurance injury, with an occupational incidence rate of 20.6[1] for disorders associated with repeated trauma in the Finance industry, which includes insurance and real estate. Only one other industry has a higher occupational incidence rate in this category: Manufacturing.

This case study was developed from a specific function within an insurance company's regional operational office. This particular function dealt with auto insurance policies, and the assembly of those policies prior to mailing to the customer. With all of the technology developed for industry, most insurance companies today still have some manual assembly of their policies before they reach the customer.

1.1 Background

Policy assembly is a typical function found within insurance company operations. Customers (policyholders) purchase coverage from an outlet, either an agent or directly via the Internet or toll-free telephone number.

Each state is charged with governing insurance; hence, policies vary from state to state. Insurance policies generally start out as a basic contract; however, they are modified to reflect the laws governing their issuance dependent upon which state the policy is to offer coverage.

Once a policyholder has been underwritten and accepted by the insurance company, a policy is issued. The mailed policy customers receive contain numerous coverage modifying clauses, known as endorsements, and language clarifying slips of paper, known as inserts, to draw attention to important coverage

---

[1] Source: National Safety Council Injury Facts, 1999 Edition

matters. It is the assembly of the endorsements and inserts that will be the focus of this case study.

1.2 Situation

In this particular regional operation, the insurance company had a unit of 14 employees and 1 supervisor tasked with assembling auto insurance policies. This was the final step prior to mailing the policy to the customer.

Repetitive motion injuries, as diagnosed by treating physicians, were significantly impacting both productivity and Workers' Compensation costs. The unit had experienced several RMI's over a 4-year period, resulting in a total expenditure of approximately $267,000 in claims costs and representing 28.5% of the unit's employees who experienced a reported RMI.

Of the 28.5% who experienced a RMI, half were returned to the unit with limited duty while the other half were placed in other jobs.

1.3 Analysis

Analysis of the situation included review of company injury records pertaining to the unit in question, determination of current unit production goals, and workstation process observation.

Review of company injury records was conducted to gain an initial understanding of the situation, and establish an area in which to focus observation. This review yielded the information discussed above, but did not reveal any specific task exposure of the auto policy assembler that lead to the resulting RMI experience.

Determination of current unit production goals is important to gain an understanding of work system pressures that may encourage worker short-cutting and risk taking. Each employee is expected to assemble a total of 600 policies per day. The average policy requires 11 documents to assemble.

Workstation process observation was conducted to determine the employee's role in the work process and identify RMI injury exposures. The assembly process starts when the policy face sheets are picked up from an "in" bin. The face sheets are reviewed for the coverage's desired, then the appropriate documents are pulled from their holding bins. The document holding bins are horizontal mail sorters measuring 56" W x 28" H x 15" D. These are placed at the back of the desk, creating a reach of approximately 22 ½", depending upon desk set-up. Once assembled, the entire document packet is stapled and placed in an "out" bin for collection and mailing.

Observations revealed shoulder flexion, elbow extension and forward back bending from extended reaches, hand and arm pronation from assembly document holder design (horizontal bins), and elevated legs and feet from seated positions elevated to accommodate non–adjustable desks.

One last item to consider is that the unit will be shut down in 15 months. This is part of an overall effort for the insurance company to consolidate operations for efficiency, which means current unit functions will be absorbed into other units, without adding staff. Production performance expectations are to remain at 600 policies per day for each employee in the absorbing unit.

2   PROBLEM SOLVING PROCESS

The first review of the unit in action was to determine the nature and scope of the unit's operation from first hand observation. Discussion with the supervisor included productivity expectation and issues, expected and planned for organizational changes that would impact the unit, and the similarity of the workstations and tasks.

From this discussion, a partnership commitment was entered into to provide time to complete workstation evaluations, to be flexible with solution ideas and employee input on those ideas, and to track solution costs.

The purpose of gaining this commitment was to allow for a solution development structure that would be conducive to evaluating the situation from an injured employee's perspective while gaining participatory input for solution development from all employees. This would increase buy–in for solution success.

2.1 Solution Development Structure

With the problem solving process purpose in mind, a multi–step solution development structure evolved.

The first step involved conducting workstation process evaluations to note conditions that lead to injury exposing postures, determine what those postures are, and spark discussion with employees regarding solution ideas. Since all workstation tasks are the same, only three workstations were evaluated. Of the 3 workstations evaluated, 2 were of employees experiencing RMI's, and 1 from an employee who had not experienced RMI symptoms. Slides of the workstations evaluated were taken for use in evaluation and unit training.

Next, training was conducted to acquaint all unit employees and the supervisor on pertinent factors and issues that lead to RMI's, and to spark thought and discussion on solutions to prevent these types of injuries. Training content was developed and the training conducted 3 weeks following the workstation evaluations. The training included anatomy, function and resultant injuries to provide participants with foundation knowledge, then proposed solution ideas obtained from the evaluations. The training class took 3 hours to complete. A 1-week period followed the training, which was designed to allow employees to think about solution ideas, before discussion ensued regarding solution ideas for implementation.

From the discussion, solutions were implemented on a trial basis. A suitable office supplier with a willingness to provide equipment on a trial basis was found. Two workstations were chosen (1 with an injured employee and 1 with an employee who had not noted any symptoms) to test solution ideas on a trial basis. Trials lasted 3 weeks before equipment was purchased for the entire unit.

Once the correct equipment was determined from trials, changes were implemented unit–wide. Each workstation was specifically fitted for the employee in a process that lasted two days.

2.2 Solution Ideas

Solutions implemented included

a. Seat adjustments to encourage neutral and supported seated postures. Adjustments were made to the seat height and backrest. For employees whose elevated chair height could not be accommodated, footrests were obtained.
b. Replacement of the horizontal document bins with 30-degree angled sorters. This was completed to minimize pronation and reduce pinch forces needed to separate individual documents required in assembly. The 30-degree angled sorters were purchased in sections of 6 (each desk equipped with 3 sections) to allow for flexible desk placement to reduce reaches.
c. Core policy documents were pre-assembled to reduce repetition during assembly. It was discovered that 5 documents were common to all policies, and could be pre-sorted into packets directly from computer generation.
d. Rearrangement of equipment on the desk surface to make workflow efficient. The most significant rearrangements involved placing the assembly document sorters within 12" of the desk front and the automatic staplers near the "out" bin. These reduced reaches, forward back bending and eliminated cross–reaches for stapling.
e. The work process was also staggered for assembly of like policy coverages at one time. Work staggering allowed for natural workflow rest breaks to occur, while reducing the amount of required assembly documents to reach over during specific assembly functions.
f. Lastly, an exercise program was implemented by a physical therapist to increase blood flow for sedentary job positions. Exercises were specifically designed as non–strenuous. Exercises took place 4 times daily, twice in the morning and twice in the afternoon.

2.3 Alternative Solutions

Two alternative solutions were raised; automation of the policy assembly process, and utilizing a slant board for face sheet review and policy assembly.

The first alternative, automation, is based upon the premise of policy endorsements and inserts coded in the computer. Once underwriting has been completed and coverages determined, the computer would automatically select, print, collate and staple the appropriate policy packets for mailing. This alternative solution was not considered since company data programming priorities circumvented this process as an option. The idea, however, was placed on a 5-year implementation program. As it turned out, automation was partially achieved by pre-assembly of core policy documents.

The second alternative, utilization of a slant board, is suggested as an additional measure of reducing reaches and forward back and neck bending for individual employees who need it, without giving up work surface. Some employees were accommodated with this solution.

3   RESULTS

All solution ideas generated were implemented, except automation. Solutions were chosen because they were considered "reasonable" in both cost and time impact on productivity.

3.1 Costs

Costs of implementing solutions were tracked. Total equipment expenditure was $2,509.00. Cost of training and employee interaction (non-productive) time was estimated to be $3,260.00, based on unit time for training and individual employee interaction time for adjustments.

3.2 Impact

Following implementation, monitoring of the unit continued for the purpose of addressing concerns raised and making enhancing adjustments to reduce exposures. The monitoring process began on a weekly basis for a period of 2 weeks, bi–weekly for 2 months, monthly for 3 months, and lastly, bi–monthly for 5 months.

The auto policy assembly unit continued to operate for 19 months, 4 months beyond scheduled operations cessation time. The unit remained fully staffed and experienced zero injuries of any type following this intervention.

During the monitoring time period, productivity gains or losses were to be measured. Productivity gains rose to an average of 668 assembled policies per day following this intervention. The unit realized a sustained productivity of 68 policies above goal, or 11% increased productivity.

4   PROCESS CRITIQUE

Feedback regarding problem approach and solution process was gathered throughout. Feedback was positive, particularly in gaining employee input regarding physical and system barriers, and involvement in the solution process. The consultant acted as a guide, providing credence to ideas of all involved and fostering an environment conducive for input.

POTPOURRI

CHAPTER TWENTY FIVE

The Economics of Ergonomics: Three Workplace Design Case Studies

Stephen Jenkins, M.Sc.
Workplace Health, Safety and Compensation Commission,
New Brunswick, Canada

Jeremy Rickards, P.Eng.
Institute of Biomedical Engineering, University of New Brunswick,
New Brunswick, Canada

Over the past three years, the authors have been engaged in a research project that has the aim of creating a methodology to identify, pre-intervention the costs and benefits, and the return-on-investment of workplace ergonomics intervention. In a paper presented at the ACE 30th Annual Conference, it was reported that the Ergonomics Audit Tool (EAT) proved to be robust, but that the design of the Cost Audit Tool (CAT) had to be re-evaluated since a number of problems were exposed with respect to the capture of cost accounting data (Burrows, *et al.* 1998).

While these modifications were introduced to the CAT, and applied in test cases, the process proved to be cumbersome, and the authors remained skeptical that the methodology was effectively representing the true production costs of a specific intervention. It was, therefore, decided to examine in much greater detail how businesses structured their cost data, and, therefore, what effect this had on the financial justification for an ergonomics intervention.

In traditional accounting systems, costs of production such as design, transportation, marketing, sales, engineering, worker compensation premiums, and health and safety, are considered an overhead (fixed) cost and are allocated, during the annual budget process, on the basis of total units produced, production floor area utilized, or some similar fixed arbitrary unit. Therefore, the actual variable production costs of an individual product or process cannot be accurately captured. As a result, overhead costs can end up being five to ten times that of direct labour costs. Hence, one particular production unit may have overhead costs several times greater than other units in the organization, but this discrepancy would be unrecognizable, and would understate actual total costs in that unit, but spread additional overhead costs onto other units.

We have examined two approaches to overcoming these accounting deficiencies. The first is Activity Based Costing (ABC), not a new technique, but one that is not in common use, especially among smaller businesses. ABC focuses on determining the actual cost of production by allocating overhead costs to the actual production function that uses them. In its purest form, it identifies all costs (sales; transportation; design; marketing; maintenance; compensation; health and safety; etc.) to a specific product or process, in effect, changing overhead costs into variable production costs. Therefore, ABC gives managers a tool by which they can identify, and take action on, all costs, not just those of direct labour and material.

The second is what we have described as Defective Production Costs and Sub-Optimal Productivity. These two terms collectively describe the cost difference between the theoretically best production process and the actual (current) production process. Engineering this difference to reduce production costs to a theoretical best is given the term A delta T (AdT). Japanese managers use the term Kaizan to describe a similar approach.

Ergonomics considers the human operators' skills and capabilities thru task analysis and job design in order to optimize human performance. Ergonomics is most usually associated with injury prevention, but a poorly designed human/machine interface can also result in sub-optimal production costs and material waste.

There is another problem arising from the affiliation of ergonomics with injury prevention, health and safety. Money spent on these perceived regulatory requirements are considered negative investments, and therefore, every attempt is made to minimize them. Hence, it is often difficult for practitioners to justify expenditures for ergonomics, and consequently, production costs and the associated losses also remain sub-optimized. The following case studies illustrate the application of both ABC and AdT, and demonstrate that this combined approach provides a powerful tool toward the goal of overcoming management's reluctance to authorize investments in health, safety and ergonomics interventions.

These three Case Studies are all taken from ergonomics interventions completed in a food processing company. They are illustrative of the approach used to arrive at a ROI and Benefit/Cost Ratio. While they are all taken from one industry, the authors' experiences are that they constitute fairly typical results of what could be expected from any small to medium size business that employs conventional accounting systems.

1. SHRINK WRAPPING OF A PROCESSED PRODUCT

1.1 Job Description:

Product is decanted from a floor-level bin, lifted onto a conveyor, placed in bags and shrink-wrapped for storage.

1.2 Ergonomic Risk Factors

Operators bending from the waist and reaching forward with the arms, below knee level. Prolonged and repetitive bending of the back, and of the upper body extremities.

1.3 AdT evaluation

Reaching to decant product at floor level required 130% of the optimal cycle time. 2% of the product is spilled and wasted on the floor. Cleaning the bin after each product cycle takes four minutes.

1.4 Workplace Redesign

The bin was raised to the worksurface height and tilted forward so that the product could slide directly onto the conveyor. Waste was eliminated. Cleaning of the bin was reduced to two minutes, a 50% reduction. Overall cycle time was reduced by 30%.

1.5 Comfort Scores

Reduced from a group average of 6.8 to 3.6 (scale 1-10).

1.6 Production Cost Savings

$293,700 per annum

1.7 Workplace Redesign Investment

$14,200

1.8 ROI / B/C Ratio / Payback period

2,100% / 20.1 / 2.5 weeks

2. PACKAGING PRODUCT FOR QUICK-FREEZE

2.1 Job Description:

Folding a box by hand, inserting a plastic liner, filling the box with product from a chute to approx. 30 lbs., weighing, strapping and palletizing the box.

2.2 Ergonomic Risk Factors

The feed chute at ankle-height requiring the operator to bend and then lift the filled box. Repetitive bending, twisting motions and increased fatigue due to the positioning of the scale and pallets.

2.3 AdT evaluation

Working close to floor height adds to the cycle time of filling and lifting the boxes. Accuracy errors at the feed chute result in loss of product which falls to the floor.

2.4 Workplace Redesign

The feeder and conveyor were raised to elbow height, eliminating bending, and reducing the cycle time by 12%. A feeder guide was installed to fill the boxes, eliminating product loss. The workspace at the box-folding table, feeder, scales and pallet was reorganized, to reduce walking.

2.5 Comfort Scores

Reduced from a group average of 7.0 to 2.8 (scale 1-10).

2.6 Production Cost Savings

$37,500 per annum.

2.7 Workplace Redesign Investment

$4,600

2.8 ROI / B/C Ratio / Payback period

812% / 8:1 / 6.5 weeks.

3. SORTING AND PALLETIZING FINISHED PRODUCT

3.1 Job Description:

Workers sort, palletize and then store more than 100 different boxed products, which vary in weight from 12 lbs. to 144 lbs., depending on the product being produced .

3.2 Ergonomic Risk Factors

Workers handle between 500 and 1000 boxes of varying weights per shift, walking >4 kms. on concrete floors. Frequent back, leg and upper body injuries.

3.3 AdT evaluation

Erratic pallet placement is caused by lack of information on the changeover of product. Increased cycle time per box results since the conveyor length is limited to12 ft., but the working area is 100 ft. long. Operator fatigue can cause the heavier boxes to be dropped, resulting in loss of product.

3.4 Workplace Redesign

The conveyor was extended in length to 100 ft. Pallet locations were standardized and marked. Storage racking was widened to improve access to pallets. Improved intercommunication with production to identify product. Increase in headroom clearance.

3.5 Comfort Scores

Reduced from a group average of 8.6 to 4.5 (scale 1-10).

3.6 Production Cost Savings

$492,400

3.7 Workplace Redesign Investment

$8,300

3.8 ROI / B/C Ratio / Payback period

6000% / 59:1 / 4 days

Ergonomics is still often viewed by management as a means to prevent injuries, while providing no return on investment. This mentality serves to hide the potential for ergonomics not only to prevent injury, but also to improve labour efficiency and reduce the cost of production. Applying ABC to these case studies resulted in increasing variable costs from 46% to 69%. However, both the actual production costs and the workplace ergonomic deficiencies where now correctly identified. Applying AdT techniques pinpointed the key workplace areas requiring re-design, leading to management action and the ROI reported in each of these case studies.

If cost accountants could be encouraged to sacrifice financial simplicity and consistency to achieve greater production cost accuracy, then it would be easier to demonstrate and justify investments in ergonomics interventions.

Burrows, E., Jenkins, S., Thomas, G. and Rickards, J. 1998, A Pre-intervention Benefit/Cost Methodology - Refining the Cost Audit Process. Proceedings of the 30th Annual Conference - Association of Canadian Ergonomists, pp. 131- 136.

CHAPTER TWENTY SIX

Using NCCI Rating Worksheet in Cost/Benefit Analysis

Subhash C. Vaidya, MS, CPE, CSP, ARM and
Arlo L. Weeks, MBA, CSP

1.1 BACKGROUND

By law every employer is required to provide workers' compensation coverage. With the exception of a few states such as Ohio, Washington, North Dakota, etc., that have the state program, and the self-insured companies, this coverage is primarily provided through private insurance carriers. Like the insurance for personal auto, the premium for workers' compensation is based upon the experience of the individual employer relative to similar businesses within the state. This experience is indicated by an experience modification rating factor assigned to the employer by a rating agency such as the National Council on Compensation Insurance, Inc., or NCCI. At the time of this writing 40 states participate in the 'experience rating' from NCCI. The rating is used 'as is', or a state version of it with some variation, to determine the modification rating factor and to more accurately predict future workers' compensation losses. A rating of 1.00 indicates average loss experience, anything below 1.00 shows relatively fewer losses, and vice versa. Insurance companies use the experience modification rating (MOD) to modify the workers' compensation coverage premium. A sample NCCI rating worksheet is shown in Figure 1. To see the explanation of experience rating worksheet components, one can check the NCCI website at 'www.ncci.com'.

Ergonomists and risk management professionals use workers' compensation data as one of the sources for identifying potential problem areas and for cost/benefit analysis. Most acknowledge and understand that past claims have an impact on the current premium, but very few have the means and method to separate it. We used the NCCI rating worksheet and the payroll based premium to separate this impact to show employers the contribution of previous workers' compensation losses in terms of additional dollars charged in premium.

1.2 METHOD

One glance at the rating worksheet shows that the past claims do impact the final rating factor. Many customers feel that a rating factor of 1.00 or below indicates an average or superior claims experience, relative to similar businesses in the state. We felt that the customer should focus on what they can do to be the 'best in business' rather than being satisfied with average. When the impact of past claims used in the rating worksheet and the additional dollar amount paid in insurance premium was clearly shown, it became obvious that except for severe losses, the employer ultimately pays its own claims.

The simplest and quickest way to compute impact of past claims is to assume no losses for the three-year period in question, i.e. columns 7 through 10 of the rating worksheet do not include any data and the numbers in the boxes at the bottom of the worksheet for actual claims turns out to be zero. This will correspondingly reduce the modification factor (Figure 1). The difference in the factor given in the rating worksheet and the factor computed under no claims scenario can be treated as the added contribution of claims to the experience rating. The modified premium (calculated as a product of the subject premium times the modification rating factor, subject premium being the payroll based premium) can then be rearranged in the same proportion into two components, revealing impact of claims and the resulting additional dollars for the current policy year (Figure 2). Taking it a step further, the contribution of each individual claim to the current policy year premium can be measured (Figure 3).

1.3 DISCUSSION

Everyone agrees that "What gets measured, gets done". Businesses have used several methods to account for workers' compensation losses that range from using actual claims related dollars to using a 'magic number' as a multiple to estimate indirect expenses such as lost time, replacement, training, claim monitoring, accommodation, etc. We wanted to identify a method of establishing a useable relationship between operating decisions and claims expenses. The results shown in Figure 3 can be used retrospectively for budgeting, and prospectively to justify process improvements.

Very few businesses go to the effort of identifying and allocating insurance expense to the individual process, department, operation, product or service. Sometimes this is acceptable because insurance expense is not a concern compared to other issues and is usually absorbed as a G&A item, but more often it is a result of a lack of better means for allocation. Using each claim as a line item, one can add up the cost of insurance premium into a fixed portion (based on payroll) and a variable portion (Figure 3). The example used clearly shows more than half of the premium is a result of previous claims, and that amount can be treated as a 'variable' and; hence, a 'controllable' expense.

Typically, NCCI issues the rating worksheet about six months prior to the renewal of the policy. The rating method uses a three-year period going back five years from the current policy period, i.e., a policy beginning in the year 2000 will take into consideration the reported losses for the years 1996 through 1998, as the numbers for the year 1999 would not be available. So when one is budgeting for the insurance expense, it can be allocated to an individual, process, department, product, cost center, etc., depending upon the budget system.

The benefit of measuring insurance expense in this manner is that the person in charge of the process, department, product, etc., gets measured for the claims resulting from the part of business they have control over and may find a greater incentive to do something about it. It is quite true that using the NCCI rating leaves no control for the upcoming and following year or two, as the rating is retrospective in nature. But knowing what it is costing at this time can help identify and prioritize target areas for improvements. Further return from such improvements should include possible savings in the premium.

We estimated the impact of operational changes on risk and on future claims related expenses under reduced risk conditions. This can be taken a step further to estimate future premium dollar savings (Figure 4). Managers can then make better informed decisions as to which projects generate greater benefit for the organization.

1.4 ACKNOWLEDGEMENTS

Authors wish to acknowledge James L. McKenzie, PhD; Laurie T. Hoskins, BS, FLMI; Monte W. Ball, BS, CSP; and Norman H. Anderson, CPCU, CSP, for their contributions to this article.

1.5 REFERENCES

The NCCI website at www.ncci.com

Internal company documents (with names and any other client identification removed) such as actual rating worksheet, policy statement, proposal document.

Figure 1: Sample Information from NCCI rating worksheet

NCCI LOGO

NAME   RISK IDENT. NO   999999999
OF RISK   STATE   AB

1	2	3	4	5	6	7	8		9	10
CLASS CODE	ELR	D RATIO	PAYROLL	EXPECTED LOSSES	EXP PRIM LOSSES	CLAIM DATA	IJ	O F	ACT INC LOSSES	ACT PRIM LOSSES
CARRIER 12345			POLICY NO. 9X999999			EFF DATE 04/01/95			EXP DATE 04/01/96	
1111	231	25	390836	9028	2257	1234567	4	F	9772	5000
2222	146	30	41614	608	182	9876543	4	F	11423	5000
3333	189	28	603405	11404	3193	1111111	4	F	14565	5000
4444	98	29	747244	7323	2124	2222222	4	F	25531	5000
5555	261	30	1119531	29220	8766	3333333	4	F	40533	5000
6666	109	29	1269722	13840	4014	4444444	5	F	3890	3890
7777	112	31	307359	3442	1067	5555555	4	F	4009	4009
8888	13	33	1565887	2036	672	6666666	5	O	4828	4828
9999	12	29	24431	29	9	NO. 5	5	*	2887	2887
1122	109	33	356455	3885	1282	NO. 43	6	*	9823	2947
2233	72	33	64484	464	153					
3344	45	33	118327	532	176					
4455	189	30	377724	7139	2142					
5566	145	31	42663	619	192					
6677	429	27	450544	19328	5219					
7788	98	31	556803	5290	1640					
9999										
POLICY TOTAL			8037029						127261	

(A) (W) 045	(B)	(C) (D-E) 253931	(D) 356977	(E) 103046	(F) (H-I) 221772	(G) 42000	(H) 320841	(I) 99069

	PRIMARY LOSSES	STABILIZING VALUE	RATABLE EXCESS	TOTALS	EXP. MOD
ACTUAL	99069	181662	99797	380529	
EXPECTED	103046	181662	114269	398977	0.95

Exp. MOD with 'zero' or no losses:

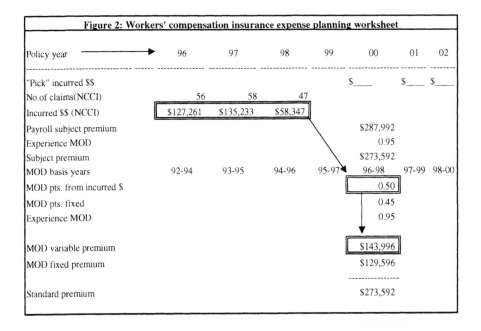

Figure 2: Workers' compensation insurance expense planning worksheet							
Policy year	96	97	98	99	00	01	02
"Pick" incurred $$					$___	$___	$___
No.of claims(NCCI)	56	58	47				
Incurred $$ (NCCI)	$127,261	$135,233	$58,347				
Payroll subject premium					$287,992		
Experience MOD					0.95		
Subject premium					$273,592		
MOD basis years	92-94	93-95	94-96	95-97	96-98	97-99	98-00
MOD pts. from incurred $					0.50		
MOD pts. fixed					0.45		
Experience MOD					0.95		
MOD variable premium					$143,996		
MOD fixed premium					$129,596		
Standard premium					$273,592		

Figure 3: Financial analysis of workers' compensation calaims

Experience MOD	0.95	Subject Premium	$287,992			
MOD Points from Incurred "$"	0.50	Premium from Incurred "$"	$143,996			
MOD Points fixed	0.45	Fixed Premium	$129,596			

CLAIM NO.	ACTUAL INCURRED LOSSES	EFFECT ON MOD	EFFECT ON PREMIUM	DEPT.	PROCESS	PRODUCT ACTIVITY	RISK FACTOR
1234567	$9,772	0.02	$5,159	Assembly	Engines	A	
9876543	$11,423	0.02	$5,695	Welding	Chasis	B	
1111111	$14,565	0.02	$6,716	Purchasing	QC	B	
2222222	$25,531	0.04	$10,278	Store	Inventory	A	
3333333	$40,533	0.05	$15,151	R&D	Modelling	D	
4444444	$3,890	0.01	$2,808	Accounting	Qtr. Report	A	
5555555	$4,009	0.01	$2,894	Warehouse	Inspection	B	
6666666	$4,828	0.01	$3,485	R&D	Training	A	
NO. 5	$2,887	0.01	$2,084	Purchasing	Misc.	Misc.	
NO. 43	$9,823	0.02	$4,361	Assembly	Misc.	Misc.	
7777777	$5,088	0.01	$3,638	Assembly	Painting	C	
8888888	$19,264	0.03	$8,242	Store	Package	A	
9999999	$27,990	0.04	$11,077	Assembly	Drivetrain	B	
1111222	$29,942	0.04	$11,711	Warehouse	Bar coding	D	
2222333	$33,682	0.04	$12,926	Assembly	Painting	D	
3333444	$8,180	0.02	$4,642	Welding	Doors	B	
NO. 4	$4,688	0.01	$3,384	Accounting	Misc.	Misc.	
NO. 48	$6,399	0.01	$2,841	Assembly	Misc.	Misc.	
4444555	$26,941	0.04	$10,736	Purchasing	Pricing	C	
5555666	$4,000	0.01	$2,887	Store	Inventory	B	
6666777	$11,071	0.02	$5,581	Welding	Chasis	A	
7777888	$2,978	0.00	$1,322	Assembly	Final check	A	
8888999	$4,145	0.01	$1,840	Assembly	Package	C	
NO. 42	$9,212	0.01	$4,089	Assembly	Misc.	Misc.	
		0.50	$143,996				

Figure 4: Estimation of premium savings related to proposed operational changes

Following estimates were presented to a client back in late 1995.

Management estimated a reduction in incurred dollars to be as follows:
Year 2 - 5%, Year 3 - 15% , Year 4 - 25%, Year 5 - 50% (Year 1 is already accounted for in MOD calculations)

The incurred dollars used to estimate future MOD are as follows:

Year 92 - $59,486 - as given in current NCCI rating worksheet;
Year 93 - $59,229 - as given in current NCCI rating worksheet;
Year 94 - $51,894 - company records,
                    1 @ $25,000, 1 @ $10,000, and several totaling
$16,894;
Year 95 - $37,537 -   1 @ $20,000, 1 @ $10,000, and several totaling $7,537;
Year 96 - $33,586 -   1 @ $20,000, 1 @ $10,000, and several totaling $3,586;
Year 97 - $29,635 -   1 @ $15,000, 1 @ $10,000, and several totaling $4,635;
Year 98 - $19,756 -   1 @ $10,000, 1 @ $5,000, and several totaling $4,756

Expected resulting MOD and premium are (considering payroll changes, medical costs, inflation, etc.):

Year	MOD	Premium	MOD base years
1996	1.10	$101,471*	1992-1994
1997	1.08	$99,626*	1993-1995
1998	1.00	$92,247*	1994-1996
1999	0.92	$84,867*	1995-1997
2000	0.88	$81,177*	1996-1998

* - these are estimates only and need to be revised as soon as new data is available to reflect rate changes, incurred dollar changes, additional process changes, etc.

The future premium savings can be expressed as and added to the project NPV. (Actual MOD turned out to be 1.00 for 1998, 0.81 for 1999 and 0.70 for 2000)

CHAPTER TWENTY SEVEN

The Power Zone

Thomas Hilgen, MS, CPE
Marsh Risk Consulting, San Francisco, CA

ABSTRACT

Ergonomics practitioners use a variety of analysis methods and measurements. When the analysis and findings are presented or reported, an ergonomist can use a system of complex terminology that the average layperson may not adequately comprehend. When the analysis or findings cannot be fully comprehended by the customer, the Power Zone concept can be used for simplification purposes. The Power Zone is a simple, yet very effective, ergonomics concept. When work is performed in this zone, the worker can use his/her large and strong muscles to perform the work with maximum efficiency. When work is placed outside of the Power Zone, the worker is in a weaker position which can increase the potential for injury. The concept of the Power Zone is discussed and its integration in workstation design and training at several companies in different industries is presented.

INTRODUCTION

Ergonomics practitioners normally use a specific analysis, presentation, and report writing method targeted to the individual "customer." The customer can range from an employee or co-worker with an MSD (musculoskeletal disorder) injury to the president of the company. Other customers include engineers, human resources professionals, safety professionals, risk managers, union representatives, plant managers, middle management, and supervisors or lead persons.

Depending on the customer, the ergonomics practitioner may use a different analysis, presentation, and/or training method. For example, when working with engineers, ergonomists normally use detailed analysis methods. When presenting possible solutions to senior management personnel, ergonomists usually summarize the major findings and include a cost-related and/or return-on-investment analysis.

Many of these ergonomics methodologies have successfully assisted customers in reducing costly MSDs, improving the quality of work, and improving worker efficiency. This chapter introduces the concept of the Power Zone which can be used as an additional tool in the application of industrial ergonomics.

ERGO-SPEAK

With any audience, an ergonomist can use excessive "ergo-speak" which may confuse or alienate the customer. Common ergo-speak includes terminology such as shoulder abduction, 5th percentile knuckle height, neutral posture, and ulnar deviations. These terms may be meaningless to the average layperson. When ergo-speak begins, many customers stop listening. The straightforward concept of the Power Zone can help the ergonomics practitioner avoid the trap of ergo-speak.

When working with a skeptical or cynical customer, the use of ergo-speak can hurt the credibility for ergonomics. Ergonomists are sometimes challenged by customers who cannot accept soft tissue injuries as real or ergonomics as an actual science. Under this type of adversity, the ergonomics practitioner can use the concept of the Power Zone to his or her benefit.

THE POWER ZONE CONCEPT

The Power Zone is a simple, yet very effective, concept. It can be used successfully for any audience in any industry with quick understanding and acceptance. The Power Zone is a simplistic version of standing and sitting workstation design guidelines (Grandjean, 1993) which are based on anthropometric data.

The Power Zone concept is the design of work and the performance of work within a specific region of which tasks can be performed most efficiently. As shown in the overlaid box in Figure 1, the Power Zone ranges from the employee's mid-thigh to the chest. It is the center of the baseball strike zone in which a player can effectively hit the ball. The Power Zone can also be compared to the zone at which weight lifters, offensive lineman in football, boxing professionals, and martial arts experts generate the most power, control, and command.

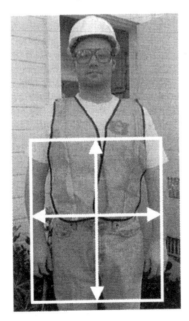

Figure 1 The Power Zone

Inside the Power Zone, the hands are positioned between the mid-thigh and chest region, the employee's elbows are in adjacent to the torso, and the upper torso is upright without bending, reaching, leaning, or twisting. Ideally, the hands are positioned so that the thumbs are forward or upward from the fingers as in a handshake position (also known as the power grip within the ergonomics community). When work is performed within each individual's Power Zone, one uses his/her strong arm, leg, and upper torso muscles to perform the work with a minimum risk of injury. This can lead to increased workplace efficiency, reduced employee fatigue, and improved work quality and production.

When work is placed outside of the Power Zone, the worker's muscles are placed in weaker position, which can increase the potential for an MSD injury. This can lead to increased worker fatigue and negatively affect quality and production.

A crucial aspect of the Power Zone concept is the proper performance of work by the employee. To help accomplish this endeavor, the workers need to be educated. Unsuccessful training sessions instruct the audience what to do or what not to do. Successful trainers educate the participants in subjects such as the spine, the discs, and the shoulders and discuss how each is negatively affected by forward bending of the torso, twisting, leaning, and reaching.

Most employees are interested in learning how their body operates. After clearly understanding that they can lift and move loads most efficiently by using the strong leg and arm muscles and maintaining the Power Zone, employees can then consciously and repeatedly apply the principles of proper lifting.

To design and perform lifting and other arduous tasks within the Power Zone, items should ideally be positioned no lower than the employee's mid-thigh level and no higher than chest height. Items at the mid-thigh level require the worker to slightly bend his/her knees to allow the strong leg muscles to help perform the work. Keeping the items no lower than mid-thigh level avoids forward torso bending to decrease the potential for back injuries. Keeping the items no higher than chest height reduces the potential for shoulder strain.

The challenges that safety and health professionals face in the real world include how to raise or lower items (or workers) so that a majority of work can be performed inside the Power Zone. Height-adjustable unitizers and pallet jacks, scissor lifts, custom platforms, and other equipment are available to raise items. Reducing stacking and storage heights and using material handling equipment can eliminate or limit lifting above the mid-chest region.

To keep the load close to the worker, unitizers with rotating platforms are widely used. Cutouts in work surfaces and eliminating any unused spaced between the employee and the item represent possible solutions. The employees can also slide the item as close as possible to himself/herself prior to performing the lift.

When low-level work is unavoidable, a person can bend the knees, kneel (with knee pads if necessary), or sit at a lower height to lower themselves and ensure that the work is performed in a modified Power Zone.

For stationary sitting and standing tasks, the primary work area should be centered about the employee's elbow height (in the heart of the Power Zone). Light assembly, data entry, and mousing tasks should be performed at about elbow height. Heavier work is normally best performed below elbow height and precision tasks (such as fine assembly work) should be performed above elbow height.

For dynamic tasks such as order picking, the most frequently picked items should be centered in the Power Zone. The less frequently retrieved items should be located above and below the central location with the heavier items below and the lightest items above the center.

UNIVERSAL APPLICATION AND ACCEPTANCE

The Power Zone concept is universally accepted. Employees, all levels of management, and unions have embraced the Power Zone concept. It has been used effectively in ergonomics training, safe lifting training, behavioral safety training, behavioral safety feedback and observation, tool and equipment design, and workplace design.

The Power Zone concept is especially useful for workplaces which have not changed much over the past several decades. Whenever one hears the phrase "this is the way it's always been done," the Power Zone concept may be the best tool to help gain acceptance and credibility for ergonomics.

The Power Zone concept works successfully in industries in which employees perform tasks outdoors with constant changing workplaces. These industries include construction, agriculture, utilities, and others. Due to the variability of tasks and often working unsupervised, employees, management, union representatives, and safety professionals can sometimes focus on one simple concept to help minimize MSDs.

SPECIFIC INDUSTRY APPLICATIONS

The usual construction employee performs varying tasks in constantly changing environments. Whether the work is performed in a completely different location or the work is in progress, the working heights, the working space, the tools and the equipment are varied. Also, a majority of the work is performed below knee level and above head height well outside of the Power Zone.

Rather than just following the status quo, some construction companies have accepted the challenge to determine methods in which low level work can be raised and overhead level work can be performed into the Power Zone. Two construction industry examples are presented: one from a drywall contractor and another from a reinforcing steel contractor.

Drywall Contractor

Drywall or wallboard is normally stacked flat on the floor. When a complete board is not needed, smaller exact sized pieces are marked and cut. As shown in Figure 2, this task was performed outside of the Power Zone.

To reposition the worker inside the Power Zone, the drywall was placed on tilted carts. Now, each board is marked, cut, and lifted in an upright position (see Figure 3). This work methods change has significantly reduced the occurrence of back injuries.

Figure 2 Bending to Cut Drywall   **Figure 3** Performing Tasks in an Upright Position

Reinforcing Steel Contractor

A majority of reinforced steel is normally assembled at the ground level. Just as it has been performed for many years, this repetitive low level work requires repetitive and prolonged bending. To help reduce the workers' compensation costs associated with back injuries, this reinforcing steel contractor developed a solution towards repositioning their employees into the Power Zone.

   Their solution was to assemble the reinforced steel in sections. These sections would be positioned upright to help their employees perform a majority of their work inside the Power Zone (as shown in Figure 4).

Figure 4 Vertical Reinforced Steel Assembly

Tree Care

In the tree care industry, each tree trimming task in different. Since every tree, every limb, and the immediate surroundings are always different, the tree trimmer's workplace is never the same. The tree trimmers either climb the tree or use a bucket truck on an adjustable boom.

After climbing a tree, the tree trimmer uses a handsaw and a manual pole pruner. For safety purposes, the tree trimmer must repeatedly secure himself to the tree prior to performing any cuts. Since the locations of the tree limbs are widely scattered, the tree trimmer used to cut any limb within reach. After the employees and supervisors received the Power Zone training, the focus is to perform each cut (unless impossible) within the Power Zone.

When using the bucket truck (as shown in Figure 5), the tree trimmer adjusts the boom to position himself properly prior to cutting. Extreme caution must be used to avoid any nearby high voltage power lines. When standing in the bucket truck, the tree trimmer predominantly uses a hydraulic pole pruner which weighs 15 pounds and is 7 feet long. He also uses a chain saw and a manual pole pruner.

Instead of cutting or pruning while bending forward, overreaching, twisting, or leaning, the tree trimmers are trained and encouraged to perform every cut within the Power Zone. The fellow crewmembers, the supervisors, and the safety staff continually and positively reinforce this desired behavior.

Figure 5 Tree Trimmer Using a Pole Pruner Within The Power Zone

Landscaping

One of the many different landscaping tasks is fertilizing trees. To fertilize trees, the landscaper inserts fertilizer with an injector gun around the perimeter of the tree branches. The injector gun is attached to a hose that connects to the fertilizer tank on the truck. The injector guns includes several major parts: the footplate, the trigger, and the handle. The footplate is used to help push (with the foot) the injector tip into hard ground. The trigger is activated with a slight squeeze, and the handle is used for added strength and support of the opposite hand.

As shown in Figure 6, the landscaper is bending his torso forward while performing the fertilizing tasks. To move the fertilizing work posture into the Power Zone, the following steps were taken:

1. Increase the length of the pipe section between the handle and the foot plate (as shown in the arrow below) to allow the individual landscaper to stand upright to use the injector gun.
2. Customize each injector gun for each individual landscaper.

Figure 6 Landscaper Fertilizing Outside of The Power Zone

Manufacturing

There are many examples of employees not working in the Power Zone in manufacturing. Employees working outside the Power Zone with awkward postures represent symptoms of an inefficient process or workplace.

When ergonomics is not incorporated into the workstation design and work area layout, awkward postures are commonplace. This is especially true when assembly tasks are performed at fixed-height tables.

An example of an assembly task performed on a fixed-height table and outside of the Power Zone is shown in Figure 7. The assembly task height ranges from her standing waist level to her eye level.

As Figure 7 shows, the bent wrists, the extended arms, and the tilted head clearly depict a workstation that was designed with minimum ergonomics involvement. Without an adjustable-height work surface, this task cannot be performed with optimum efficiency or work quality since it is performed outside of the Power Zone.

Figure 7 Assembly Tasks Performed at Eye Level

CONCLUSION

The Power Zone is a simple ergonomics concept that can be used as a target for workplace design, a desired behavior, a quality initiative, a slogan, and many others. It can be used effectively in many workplaces, and the phrase and its meaning can gain universal acceptance in the workplace.

The Power Zone is not a comprehensive ergonomics solution. It is not a substitute for detailed ergonomics evaluations or the use of sophisticated analysis and measurement tools. But, as this chapter presents, the Power Zone concept can serve as a valuable tool in training and workplace design.

REFERENCE

Grandjean, E., 1993, *Fitting the Task to the Man: A Textbook of Occupational Ergonomics*, 4th ed., (London: Taylor & Francis).

CHAPTER TWENTY EIGHT

Ergonomics Programs: Reducing Work-Related Musculoskeletal Injuries

Alison G. Vredenburgh, Ph.D., CPE and Ilene B. Zackowitz, Ph.D.
Vredenburgh & Associates, Inc.

Dudley Wainright is the risk manager for Skyline Mountain Community Hospital*. Over the past six years, he has noticed an increasing rate of turnover. Furthermore, the employees' back-injury rate has tripled and the worker's compensation costs are through the roof. The employees frequently complain about under-staffing and the constant need to hurry. Despite his current ergonomics injury reduction program (EIRP), many employees are out on permanent disability. At any one time, more than 20% of the employees are new to the job, having been hired within the previous three months. Further, the hospital's insurance carrier is considering dropping the hospital's coverage due to its high injury rates. The hospital administration has just called Dudley "onto the mat" and told him he'll be terminated if the injury rate does not drop significantly within two years, when the Joint Commission on Accreditation of Healthcare Organizations (JCAHO) returns to review the hospital's progress. As with any hospital, Dudley knows that his budget is limited.

Fictitious name

Why is a "systems" approach to ergonomics program evaluation important?

What features characterize effective ergonomics programs?

What steps should Dudley take to develop a more effective ergonomics injury reduction program?

1.1 WHY IS A "SYSTEMS" APPROACH TO ERGONOMICS PROGRAM EVALUATION IMPORTANT?

An increased emphasis on research in the areas of ergonomics and safety management has been motivated by a variety of practical and theoretical concerns, including compliance with OSHA requirements, insurance costs and litigation, and a desire for increased organizational effectiveness. Consultants in the areas of ergonomics and organizational development help companies manage change efforts.

They often address issues pertaining to ergonomic injury reduction. The science of ergonomics (or human factors) is concerned with developing knowledge about human performance capabilities, limitations and other characteristics as they relate to the design of the interfaces between people and other system components. The safety and reliability of complex systems depend on failure avoidance and effective process control, both of which require a systems management approach.

Critical incidents, like occupational injuries, typically do not occur as a consequence of an isolated technical or operator error. They tend to be end-events in a poorly understood chain of interacting factors that often involve several levels within a system (Wilpert, 1994). Good programs should not only address critical problems, but also identify and resolve these problems in a timely manner before they become critical. It is restrictive to discuss failures of large-scale systems solely in terms of its technological aspects. Instead, additional factors, such as organizational level (e.g., individual, group, organization) should also be incorporated into the design, construction, operation, and monitoring of an effective ergonomics program (Vredenburgh, 1998).

When revising the ergonomic injury reduction program (EIRP), Dudley should consider adopting a systems approach, taking into account specific features of the environment in which Skyline Mountain Community Hospital operates. Within large-scale organizations, like this one, departments are inter-related and have both predictable, and unpredictable, effects on each other. Dudley must consider, for example, the hospital's laundry department, a seemingly independent segment of the hospital community. However, it is imperative that this department is considered within the system of the hospital. Many hospital employees interact with the laundry department and have an impact on laundry workers' job-related injuries. Nurses, housekeepers, and hospital techs may all overstuff and tear laundry bags, leading to excessive bending and reaching by laundry employees. The entire *system* must be evaluated. Only then will Dudley be able to develop effective interventions that reach all relevant employees.

Figure 1.1 Features of Effective Ergonomics Programs

1.2 WHAT ARE FEATURES OF EFFECTIVE ERGONOMICS PROGRAMS?

Research in applied ergonomics highlights several factors that are important in developing effective ergonomic programs (see Figure 1.1). When adopting a systems approach, these issues must be considered at every level of the organization.

1.2.1 Worker Participation

Worker participation is a term used to describe the involvement of employees in decision-making within an organization. Employees close to the work are often recognized as being best qualified to make suggestions about improvements. Participative managers solicit opinions from their employees prior to making final decisions, especially when making decisions that directly impact them. Empowering employees in this way provides them with authority, responsibility and accountability for required decisions. It also helps insure that employees and their managers jointly set goals and objectives. Returning to the opening story, Dudley should involve employees throughout the change process of the injury reduction program — from initial meetings through its implementation.

1.2.2 Ergonomics Training

In order for employees to be active participants in an injury reduction program, they must receive appropriate ergonomics training. Training provides the means for making injuries more predictable because potential injury scenarios are described and safe work procedures are recommended. Thus, through training, employees learn what tasks pose the highest risk, and how to best perform them while avoiding injury. The difference between safe employees and those who frequently get hurt is that safe employees can recognize hazards and hazardous actions and understand their consequences. Using our example, to improve the quality of safety and health for all employees, Dudley should institute a systematic, comprehensive safety and health-training program. He should then verify that the new procedures are being performed in the work areas.

1.2.3 Hiring Practices

The development of an effective ergonomics program can be facilitated if recruitment criteria for new personnel include the selection of people who are predisposed to displaying a safety-conscious attitude in their work; in other words, past behavior is the best predictor of future performance. If an organization fosters a safety-conscious image, the recruitment task will be influenced because those with compatible attitudes and expectations would be more likely to seek out this company, at least in part, because of a desire to work in a safe environment (Turner, 1991). The hiring process can be guided to ensure the proper placement of risk-conscious individuals. Dudley should assess how new employees are being selected at Skyline Hospital.

1.2.4 Reward System

Individuals are motivated to behave in ways that lead to desired consequences; they will modify their behavior if they perceive that the behavior will lead to a beneficial outcome. Incentives are used as a motivational tool in virtually all areas of business and industry; these include reward programs specifically designed to reduce injury rates. A reward system can be used to reinforce employees who call attention to ergonomic safety problems and those who are innovative in finding ways to locate, assess and remove workplace hazards (Ostrom, Wilhelmsen, & Kaplan, 1993). A key characteristic of a successful incentive program is that it receives a high level of visibility within the organization. A well-designed safety-incentive program rewards the reporting of a hazard or an unsafe act while giving bonuses for fewer lost-time accidents. Dudley should recognize that a well-designed incentive program offers recognition, which can help modify behavior of Skyline's employees.

1.2.5 Management Commitment

Management commitment has been identified as an important factor for the success of an injury reduction program. This commitment can manifest itself through job training programs, management participation in safety committees, consideration of safety in job design, and review of the pace of work. An employee's motivation to perform a job safely is a function of both the individual's own concern with safety as well as management's expressed concern for safety. For Dudley, it is imperative that he has management buy-in, from administrative executives to department heads. Without their commitment to EIRP, Dudley will face an uphill battle.

1.2.6 Communication and Feedback

The role of feedback concerning employees' performance is critical because behaviors resulting in injuries are not typically new occurrences. Their causes are deeply rooted in past minor incidents, where damage was insignificant and workers and bystanders were not injured (Kletz, 1993). Dudley can communicate regular feedback regarding performance to employees through posted charts and a review of behavioral data in safety meetings.

1.3 WHAT STEPS SHOULD DUDLEY TAKE TO DEVELOP A MORE EFFECTIVE ERGONOMICS INJURY REDUCTION PROGRAM?

After initial interviews with employees and management, it is time to develop a more effective ergonomic injury reduction program. Dudley must first identify his budget and work within this constraint. One way to decrease program costs for the hospital is to solicit involvement from the hospital's insurance carrier. Because the insurance carrier has expressed concern regarding the workers' compensation costs, it may be willing to share the costs of revamping the EIRP. This cooperative strategy will decrease the costs for the hospital and will ensure that the insurance carrier has a stake in the program's effectiveness. By reducing the costs to the hospital, a more comprehensive program will be feasible. Further, because of their own involvement, the insurance carrier is less likely to drop the hospital's coverage if they see that proactive steps for injury reduction are being instituted.

The next decision for Dudley is to determine the organizational change agent, or who will be used to guide the development of the program improvements. A qualified hospital employee is one option. The greatest benefit in using an internal hospital employee is the cost factor. This person is already employed by the hospital so their compensation will not be taken out of the program budget. However, there are several disadvantages to consider with an internal change agent. Because the person is part of the hospital system, they may have ulterior motives, preconceived notions about what is needed in the hospital, and a hand in the politics of the organization. All of these factors may interfere with developing the most effective program.

If the insurance carrier is willing to share the costs of program development, or if it is deemed within the budget, an external consultant may be the best option. Although an external consultant is more expensive, that person comes to the hospital unbiased concerning the hospital's politics, and has the experience of guiding the change process for multiple organizations. An external consultant is more able to get at the root of the issues without being swayed by extenuating factors. The change agent should use the following steps to develop the most effective ergonomic injury reduction program (see Figure 1.2).

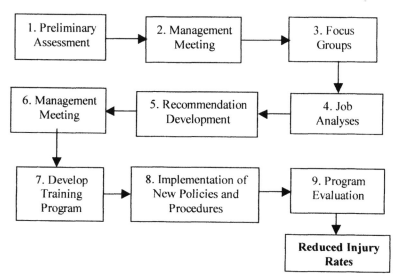

Figure 1.2 Steps to Developing an Ergonomics Program

1.3.1 Step 1: Preliminary Assessment

Before Dudley can begin thinking about ways to decrease the injury rate at the Skyline Hospital, he must begin by assessing his existing EIRP. Based on the high rate of injury, it is apparent that the current program is not effective. In order to develop a better program at the hospital, Dudley must first analyze where the present program is falling short.

Another reason why assessing the EIRP is so important is that as with most hospitals, Skyline's resources are limited. Dudley must first assess the existing program in order to allocate the available resources. By focusing on the areas in greatest need of improvement, Dudley will get "more bang for his buck."

The current EIRP should be evaluated based on the six features of effective ergonomics programs discussed above. For example, perhaps Skyline Hospital has an effective reward system, but the training and communication/feedback features are lacking. In this situation, employees may be motivated to get rewards they deem valuable, but don't know the specific behaviors that lead to the desired rewards because training is lacking. Further, the employees are frustrated because they are not getting the feedback that would illustrate what they did wrong and what they can do to achieve the rewards.

The "systems" approach is important to keep in mind at this phase of Skyline's ergonomics evaluation and program development. Data should be collected from employees at various levels and from all departments of the hospital to determine where there is agreement on the assessment of the different elements of the current program. Disagreement at this preliminary assessment phase should be considered an important signal to management. Differences in perceptions indicate areas that require further study. Perhaps in his initial assessment, Dudley found that management believes that the employees are highly involved in the EIRP because many employees attend regular safety meetings and generally have information to contribute. Dudley may be surprised to discover that there is disagreement between management and employees in this regard. Upon further investigation, he may learn that although employees attend safety meetings, their suggestions are not seriously considered or followed up on. Thus, employees do not perceive a high level of involvement. This difference in perception is imperative to consider during the program change process.

Prior to the initial meeting with management, Dudley and the change agent should review injury records to identify injury patterns for type, frequency and department of injured employees. Based on the initial assessment as well as inter-departmental follow-up, it can be determined where Skyline Hospital needs to channel its resources for maximum gain and effectiveness of its EIRP.

1.3.2 Step 2: Management Meeting

In order to ensure that the changes to the program will be effective, a preliminary management meeting is essential. Because management buy-in is vital for program follow-through, all department heads must be present at the initial management meeting. If a department is left out of this first step, it is less likely they will be committed to the revamped EIRP. Through discussion at this initial meeting, it should become evident how each department is impacted by ergonomics injuries.

The change agent will discuss why it is necessary to implement changes to the EIRP, the goals of the program change, and how the program will be developed and implemented. At this meeting, it is necessary to get commitment from all department heads that safety is an overriding priority. If this commitment is difficult for a specific department, the reasons for this problem should be discussed at the outset.

1.3.3 Step 3: Focus Groups (Meeting with knowledge workers)

The next step of program change is to identify activities and hospital procedures that lead to increased injuries. A good way to gain access to these data is through hospital employee focus groups. Because employees are closest to the work at hand, their involvement is crucial in identifying specific jobs, tasks, and procedures that lead to increased risk of injury.

Focus groups should be held for every department that has high WMSD injury rates and should include employees viewed as most knowledgeable within these departments. The focus group facilitator or program change agent should encourage employees to discuss the parts of their work that they consider most likely to lead to (or have led to) injury. In order to get more candid information, department managers may be excluded from the focus group sessions.

1.3.4 Step 4: Job Analyses

After completion of the focus group sessions, the activities should be identified that are considered most likely to lead to injury occurrence. The next step is job and task analysis. The purpose of job and task analysis is to evaluate how tasks are presently being performed. The goal is to modify tasks to be consistent with sound ergonomic practices and develop organizational policies and practices that will lead to reduced injury occurrence.

The positions to be analyzed are based on the data gathered from the focus group sessions. Job analyses should be performed on the jobs and tasks identified as having resulted in, or most likely to lead to, injury. Further, the task performance of more than one worker should be analyzed so that the data are applicable to other employees who perform that task and to identify when and why multiple approaches are being used.

The best method to collect job analysis data is through combined observation and interview techniques. Observation is essential to identify the specific behaviors that are necessary to perform the task. Interviewing the employees is usually necessary to obtain information that is not readily apparent through observation. For example, there may be underlying reasons why a hospital housekeeper carries trash to the chute in an awkward manner that cannot be determined simply by observation. A useful interviewing method, "oral protocol" is where employees describe tasks as they are performing them. This approach helps identify steps workers may not think of when just describing tasks. A dialogue between the employee and the evaluator is necessary to get all available information.

Finally, it is necessary to review the data with the employee and a member of the department's management. This is to ensure that all relevant task categories are included and that extraneous data are eliminated.

1.3.5 Step 5: Recommendation Development

Now it is time for the change agent to use the collected data to make recommendations for improvement. Based on the information gathered through the historical data, focus groups, and job analyses, new policies and procedures will be suggested to reduce or eliminate the potential risk factors that have been identified. Recommendations regarding policies include administrative controls. These policy changes must be executed by top management to ensure that implementation is successful. Recommendations regarding procedures should include employee-training issues. For example, if it was observed that employees are not lifting properly, correct lifting procedures could be demonstrated and practiced at subsequent employee training sessions.

1.3.6 Step 6: Management Meeting

After recommendations have been developed, it is time for another management meeting. This meeting will help to ensure that the proposed injury reduction solutions are feasible for a given hospital. It would be ineffective to further develop organizational policies and procedures, or to create a training program based on recommendations that hospital management considers undesirable or impossible to carry out. If a manager feels a recommended procedure is not feasible, the change agent must work with him or her to develop an approach that is more likely to address the ergonomics risk factor effectively.

A second management meeting at this point is also helpful to ensure continued buy-in and commitment. Managers must be included in the change process. Asking them for feedback *before* program change is implemented is one way to get management more involved in the injury reduction program, and ensure buy-in and support. Implementing programs without conferring with department heads will most likely reduce program effectiveness.

1.3.7 Step 7: Develop Ergonomics Training Program

The next step is to develop a training program to ensure that employees are familiar with the safest method to perform their tasks. In the healthcare setting, training will often include instructions on proper lifting procedures and demonstrating the use of assist equipment. In order to implement an effective training program, all affected employees must participate. Training should be tailored to specific departments in order that all information is relevant to employees in attendance. Further, the ergonomics training must be mandatory and on company time.

Effective training sessions are interactive so employees participate and practice the techniques. Employees may volunteer to give examples of what constitutes safe or unsafe practices. Employees may be asked to demonstrate how a mechanical lift works, or to demonstrate what they consider the most effective way to make a bed. Hands-on training is most effective, and participation from all attending makes for a salient and memorable training session.

Visual aids are another effective method for training sessions. The presentation of slides of the employees and their coworkers taken during the job analysis phase makes for memorable, and sometimes even funny training sessions.

1.3.8 Step 8: Implementation of New Policies and Procedures

At this point, the change agent should prepare a final report outlining the risk factors identified and the new policies and procedures that have been developed to address these factors. Department heads that will be affected by the changes should be given the opportunity to approve this report. At this point, the hospital management must take the initiative to verify that the recommended policies and procedures are put into practice. Performance reviews should take into account employees' safety behavior.

Feedback needs to be provided to employees, both individually and as teams or departments. Safety meetings should reinforce the new practices and discuss any problems with implementation. Posters should track WMSD injury rates by type and department on a quarterly basis. In order to evaluate the efficacy of the ergonomics program, several quarters of injury data prior to the intervention need to be evaluated.

1.3.9 Step 9: Program Evaluation

In order to ensure that the new ergonomic injury reduction program's policies, procedures, and training are effective, a program evaluation should be conducted. This will not only estimate the usefulness of program improvements, it will help satisfy accountability requirements of the program's sponsor.

Both qualitative data (through interviews, focus groups) and quantitative information (direct observations, surveys, injury data) should be included. A hospital may wish to phase in changes to one part of the hospital (one medical floor) and use another floor as a control group. Considerations should be made as to how often data should be collected (typically quarterly). It is also important to determine who will be collecting the data (e.g., a safety officer or outside consultant).

1.3.10 Back to Dudley

By following these guidelines, Dudley will be able to develop and implement an ergonomic injury reduction program that will achieve several goals. First, it will help reduce the occurrence of injuries to the hospital employees. It should decrease turnover and worker's compensation expenses. It will also help ensure that the hospital's insurance carrier will maintain their coverage. Finally, it should satisfy the JCAHO when they come for their review. And of course, his revamped program will help save Dudley's job.

Kletz, T.A., 1993, Organizations have no memory when it comes to safety: A thoughtful look at why plants don't learn from the past. *Hydrocarbon Processing,* **6**, pp. 88-95.

Ostrom, C., Wilhelmsen, O.C., and Kaplan B., 1993, Assessing safety culture. *Nuclear Safety,* **65**, pp. 163-172.

Turner, B., 1991, The development of a safety culture. *Chemistry & Industry,* **4**, pp. 241-243.

Vredenburgh, A.G., 1998, *Safety management: Which organizational factors predict hospital employee injury rates?* Doctoral dissertation. California School of Professional Psychology, San Diego, CA.

Wilpert, B., 1994, Industrial/organizational psychology and ergonomics towards more comprehensive work sciences. *Proceedings of the 12th Triennial Congress of the International Ergonomics Association,* **1**, pp. 37-40.

CHAPTER TWENTY NINE

Ergonomics Certification: Its Value and Quality

Roger L. Brauer and Valerie Rice

1.1 INTRODUCTION

The growth of titles and certifications in the United States has mushroomed in the last decade. The number of titles available for various professional fields has more than doubled since 1990. There may be a rush to have letters following one's name. For some, the more letters, the better. For some the more letters there are, the more credibility they believe they have with peers, employers, customers, clients and governments.

As the 21st century begins, there are at least nine titles available for those in the United States practicing ergonomics in some form. The titles are offered by certification boards, by schools and by private companies. Among the allied fields of ergonomics, safety, industrial hygiene and environment, there are no less than 150 titles to choose from. When titles offered outside the United States are included, the list approaches 200 titles.

The purpose of this chapter is to provide information about the certification business, the value that certification adds to the ergonomics profession, and to identify factors which help those who seek or rely on certifications to be good buyers and consumers. This section also summarizes information about the ergonomics titles available.

1.2 THE CERTIFICATION PROCESS

While there are variations in practices for particular certifications and titles, this section summarizes the general characteristics of certification programs.

1.2.1 The Business of Certification

The certification business is simple in concept. A certification organization sets standards and evaluates individuals to determine compliance with the standards. For those who meet the minimum standards, the certification organization authorizes the individuals to use a title owned by the certification organization. Typically, the standards cover education or training, experience, and demonstrated knowledge and skill through examinations. Most certifications also require periodic recertification to encourage keeping up with changes in practice.

The certification business gets complicated because it must deal with fairness, being representative of practice, having its business reviewed for financial, business and governance practices and being legally defensible.

Overall, the goal of certification is to assess competency in a particular discipline or practice area. The figure below illustrates the certification process. It is not possible to measure competency fully. People do not perform exactly as expected at all times or to the satisfaction of everyone. However, it is possible to assess with reasonable accuracy key factors which are good predictors of competent performance.

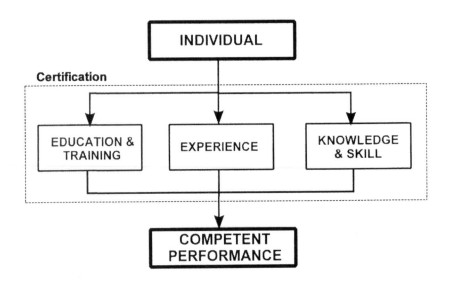

Employers try to assess competency in recruiting and hiring. Most job position descriptions include education and experience requirements. However, when reviewing applicant resumes, it is not possible to tell from these criteria alone who will perform best in an actual job setting. Employers often rely on certification as an additional evaluation of qualifications and a means to help estimate competent job performance.

1.2.2 Education and Training

Most professions require a college degree as the entry qualification. The levels of education required will vary with the certification program. For some titles, a college degree may not be an appropriate standard. Instead, there is a prescribed program of study, training or courses. In some cases, an organization may award titles after a person completes one or a series of prescribed courses of instruction. There may or may not be a capstone or series of examinations involved in the courses.

Today, people in different jobs practice ergonomics in some form. There are specialists who work only in ergonomics. For others, such as safety professionals and industrial hygienists, ergonomics has become a part of their professional practice. Ergonomics has also affected engineers and designers of furniture, equipment and workstations. Even people involved in sales may apply ergonomics. Employers often train workers to understand some aspects of ergonomics relating to their work and workstations. Ergonomics has affected many positions through regulations, standards and advertising. The level and type of training or education required for each of these will vary.

For professionals, evaluation of degrees submitted for certification credit typically relies on national accreditation of educational institutions by regional organizations recognized by the American Council on Education. There are some educational institutions in the United States which do not hold such accreditation. Degrees from these universities are usually not accepted for credit toward certification.

Certification programs may relay on program accreditation in additional to institutional accreditation. Program accreditation is typically operated by an organization aligned with the profession for which certification is offered. The Human Factors and Ergonomics Society conducts accreditation of ergonomics degrees. Ergonomics is an element in safety and industrial hygiene degrees, which are accredited through the Accreditation Board for Engineering and Surveying. Again, most universities and certifying bodies rely on recognition of the program accrediting body by the American Council on Education.

The usual procedure in the United States is to evaluate degrees from schools outside the United States for equivalency to educational standards of the United States. For example, a number of universities in other countries award bachelor's degrees after three years of college study. They are not equivalent to the four-year standard found in U.S. bachelor's degrees.

When evaluating degrees presented by certification applicants, there are two key considerations. One is verifying that the degree was awarded to the applicant. In the United States this is easily established by means of an official transcript from the cited university. The second consideration is verifying that the courses prepared the applicant for practice. Certifying bodies can check lists of accredited institutions and accredited programs published by the appropriate accrediting body. For degrees from universities outside the United States, the procedure becomes more complicated. Typically, the list of completed courses must be evaluated.

Certification programs which verify that education and training requirements are met provide a valuable service to those who rely on the certification.

1.2.3 Experience

Most certification programs have one or more experience requirements. The standards may include full-time or part-time experience or having a certain portion of a full-time position devoted to the practice for which certification is awarded.

Some certifications require a certain number of hours of practice, while others use years of practice.

One characteristic for experience standards involves breadth of practice. Standards must define the boundaries for practice to decide what experience is within the domain of practice. Experience standards must also define what mix of functions are sufficient for practice. Sometimes experience presented in an application is too narrow to be accepted.

Another key characteristic of experience involves depth. Very often professional practice experience can be distinguished from sub-professional practice by the tasks describing job duties. Some key elements that distinguish professional practice are:

- Having responsible charge.
- Making recommendations to others.
- Defending methods used to arrive at recommendations.
- Performing analysis, synthesis and problem solving for deriving recommendations.
- Being recognized or relied on as an authority for the subject area.

Practice in most professions includes compliance with laws, regulations, and standards. However, professional practice goes well beyond compliance matters. One can teach almost anyone the rules and procedures which are standard to an area of practice. One can teach people how to comply with laws, regulations, and standards. All levels of practice in a discipline may be involved in compliance matters. What distinguishes professional practice from sub-professional practice is relying on principles, theories and methods in situations for which there are no standards.

A certification program that has clearly defined experience standards and procedures for evaluating the experience of applicants provides a valuable service to those who rely on the certification to help assure competent practice.

1.2.4 Assessing Knowledge and Skill through Examinations

The third standard of certification, and the one which gets the most attention, is the certification examination and the examination procedures. Some important considerations in testing candidates to determine if they meet a minimum standard of knowledge and skill are the following:

- Deciding what subjects constitute the body of knowledge and skills required for effective practice.
- Deciding how the subjects should be organized and what portion of the examination should be devoted to each subject.
- Deciding on the examination format.
- Deciding on scoring procedures.
- Deciding on a passing score or cut score.
- Defining the testing environment.
- Determining whether the examination measures knowledge and skills effectively.

Critical to examination development and delivery is security. If the examination is compromised, it completely loses its value. If there is no examination, the certification has little value.

For most professional examinations in the United States, the preferred format is multiple-choice questions. Most professional examinations contain approximately 200 question in order to be able to achieve an acceptable reliability level. Some examinations use paper-and-pencil methods. Others use computer-based testing. A few professions with large numbers of candidates have moved to computer adaptive testing. Each delivery method has its advantages and disadvantages.

The best questions on professional examinations result from having individuals in practice participating in question development. Many quality control steps and many kinds of editing will ensure successful question performance. Editing steps include technical editing, psychometric editing and editing for grammar, style and readability.

Professional examinations are quite different from examinations used in scoring students in college courses and are quite different from achievement tests. Every question in a professional examination must have an authority for the answer, must contribute to establishing whether candidates demonstrate minimum knowledge and skills, and must be able to withstand legal challenges. In some cases, questions appear on examinations and do not count against the passing score in order to pretest their performance.

1.3 THE VALUE OF CERTIFICATION

A certification can add value to the profession, to those who practice in the profession and to those who use the knowledge and skills of practitioners. This section will review some of the values which certification provides.

1.3.1 Value for the Practitioner

For the individual practitioner, the first value that results is the personal satisfaction which stems from having achieved a standard established by peers in the profession. This may elevate self esteem and confidence and make one eligibility for work, advancement, bonuses, pay increases or increased responsibility.

A problem faced by many disciplines today is a change in the employment model. A generation ago, someone graduated from college with a degree and went to work for an employer. These people expected to stay with that employer their entire careers and relied on the employer to evaluate their job performance and advance them to higher levels of responsibility and pay.

Today that model is no longer valid. Companies have removed many layers of management and reduced professional staff positions. Some companies now contract for services which internal staff previously provided. Companies have changed structure in order to respond quickly to changes in the marketplace, to be able to incorporate new technology more quickly, and to get products designed and to market faster.

Some project that people entering the workforce today will change jobs six or more times during a career. In order to compete for positions or work, individuals must now take charge of their own qualifications. Certification provides a means for individuals to validate competency in particular areas to potential employers or when competing for work or contracts.

1.3.2 Value for the Profession

Certification can add value to a profession. The profession itself can benefit by having a means to assess minimum competency. Individuals in the profession recognize the competency of other practitioners within the profession based on the certification held.

There are many who argue that they do not need certification to be effective practitioners. They feel that their education, experience and recognition in the discipline speaks for itself. In many cases, that is true. However, certification can simply confirm that the individual is competent and increase the visibility for the claimed competency. Certification provides a means to verify competency.

One of the most frequently cited values for a certification process is that it "made me learn the subject matter." Simply going through training or education may not be sufficient. Being able to demonstrate knowledge and skills by examination requires a more thorough understanding of the subjects.

1.3.3 Value for the Employer

With the increased complexities of disciplines and specialties within disciplines, the human resource departments have difficulty in determining qualifications for positions and when evaluating candidates for employment. Certification plays an increasing role in the employment process. Human resource departments are likely to include particular certifications as qualification criteria for job applicants.

If a certification does a thorough and reliable job in evaluating education, experience and knowledge and skills by examination, those trying to screen candidates can rely on the certification to help them decide which candidate is best for their job needs. The Board of Certified Safety Professionals has monitored advertisements seeking safety professionals during the last 20 years. The portion of ads containing citations for Certified Safety Professionals (CSPs) have grown from about 20 percent in 1980 to 60 percent in 1999. The same advertisements typically include education and experience requirements about 90% of the time.

Many employers have recognized that individuals who hold certifications or licenses know their subject areas and make better leaders. The employers rely on their knowledge and skills and are willing to pay them more.

Recent data from several sources show that safety and industrial hygiene professionals who hold recognized certifications receive $15,000 per year more than similar individuals who do not hold the certifications. During a career, this pay differential can exceed $300,000.

Holding multiple certifications or licenses which are well recognized add even more pay. For example, holding both the CSP and a Professional Engineering license increases income by about $8000 per year compared to holding the CSP alone.

Holding both the CSP and the Certified Professional Ergonomist certifications increases pay by about $4000 per year.

1.3.4 Value for Government Agencies

Governments at federal, state and local levels have a responsibility to protect the health, safety and welfare of the public. Laws, regulations and standards help ensure that services offered to the public or performed under contract for the government agency provide the desired protection. More and more government agencies are requiring certifications in regulatory and contractual documents. In general, government agencies do not want to expand licensing and rely on certifications operated by peer organizations.

A dilemma for government agencies is how to cite minimum competency for a profession or practice area. Every certification organization believes that it's title should be recognized. The problem for the government agency is being able to identify those titles which reliably predict competent practice. If an agency cites one title, the others will lobby for similar recognition. As a result, there is a growing reliance by government agencies on certifications which have been independently evaluated by a third party for compliance with national certification standards. The process of accrediting certifications is becoming more important to identify quality certification programs. Section 1.4 explains the certification accreditation process further.

Another approach is to establish national standards for an area of practice which extends the quality standards for certification to a particular field. An example is ANSI E1929-98, *Standard Practice for Assessment of Certification Programs for Environmental Professionals: Accreditation Criteria.* To a great extent, this kind of standard simply restates the process used in certification accreditation.

1.3.5 Value for the Public

The public receives value when services performed for them directly by practitioners or on their behalf through government agencies are performed competently. In most cases, the public does not buy ergonomic services directly from practitioners. Employers and government agencies are more likely to purchase ergonomics services. Then, liability for ergonomics practice rests with the employer or government agency.

1.4 QUALITY IN CERTIFICATION

How can a practitioner, employer, government agency or the public determine whether the letters behind someone's name really mean anything? Do the letters really ensure minimum competency? This is a growing concern for certifications and titles in many fields. This section explains how accreditation of certifications is attempting to address this issue. This section provides only a summary of some standards for peer certification boards.

1.4.1 National Standards and National Accrediting Bodies

There are organizations in the United States which have a mission to ensure quality in certification and to accredit certification programs and organizations offering certifications. They provide a valuable service.

One of the broadest and oldest accrediting bodies is the National Commission for Certifying Agencies (NCCA). It was formed in the 1970s through a grant from the U.S. Department of Health, Education and Welfare. It sets standards for peer certifications and evaluates certification programs for compliance with the standards.

Another organization which performs these functions is the Council of Engineering and Scientific Specialty Boards (CESB). It grew out of a national symposium on credentialing in engineering and related fields in the 1980s. Its standards are very similar to those of NCCA.

In the fields of medicine and nursing, the professional model requires state licensing first and then peer certification in specialties as an option. Each of these disciplines has a governing body which sets standards for the individual specialty boards and the processes used for awarding certifications. For medicine the standard setting body is the American Board of Medical Specialties. For nursing it is the American Board of Nursing Specialties.

1.4.2 Business Standards

Accreditation standards include criteria related to finances and governance. Financial information is normally provided through annual audited records. Accreditation considers the financial health of the certification to ensure the certificate holders that the certification body will continue to exist. Another concern is knowing where the money goes to ensure that the officers and directors maintain an honest operation and are using assets wisely.

Governance criteria intend to ensure that the certification body is made up of a cross section of the practice for which certification is awarded. Governance criteria also seek to ensure that the leaders of the organization cannot serve endlessly. The criteria require that the nominating and election process allow for representation from across the profession. National standards also require public participation in certification governance by someone not involved in the profession being certified.

1.4.3 Fairness to Candidates

There are several criteria addressing fairness to candidates. For example, the certifying body must publish its procedures for appealing decisions of the governing body with regard to qualifying for, awarding or removing certification. There must be due process. Fairness includes independence between membership in an organization and the ability to pursue or hold a certification. Certification bodies are not membership societies. Fairness includes reasonable separation between preparing for and completing certification examinations. Fairness includes policies which clearly comply with non-discrimination laws and laws relating to disabilities.

1.4.4 Valid and Reliable Examinations

Accreditation criteria require valid and reliable examinations. Examination validity involves the content of examinations and what subjects they cover. There are accepted practices for defining examination contents. The subjects and their distributions are often called the examination blueprint. Typical procedures for defining certification examination blueprints involve a role delineation workshop or job analysis workshop. A panel of practitioners representing the population for which the certification is targeted get together for one to three days under the guidance of a qualified facilitator. During the workshop, the panel identifies the main functions that make up the practice and the responsibilities within each function. Then the group defines the knowledge and skills necessary for each responsibility.

Very often a second activity is included in the validation process. The results of the workshop are converted to a survey format. The survey questionnaire is sent to a significant number of people who are identified as being in the practice. The survey asks people to rate the importance, time spent on, and criticality of each function and responsibility. The survey may also address the knowledge and skills definitions. The ratings of respondents are tallied and a calculation procedure estimates the portion of the examination questions which should be devoted to each function and responsibility.

This kind of formal validation procedure ensures that the examination leading to a certification reflects what people actually do in practice. Without clearly followed procedures to define examination content, a certification examination cannot meet validation standards.

Once the examination blueprint has been established, then questions are developed and classified to meet the blueprint. When there are enough questions in the question bank to meet the needs of each subject area, the examination can be composed. The question development process must progress through several quality assurance steps, such as technical, psychometric and English editing, review by practitioners and pretesting.

Examination reliability refers to the degree to which the examination will achieve the same results when given to the same person more than once. Reliability statistics are derived from actual examination administrations. The possible range of reliability runs from zero to one. An acceptable level is usually 0.8 or higher. What some candidates do not realize is that retaking a reliable examination after failing it will result in essentially the same score unless the candidate has spent time learning the material covered on the examination.

1.4.5 Setting Examination Passing Scores

Accreditation criteria require that the method used to set the passing score must ensure that all candidates have an equal chance of passing the examination. The passing score or cut score represents the minimum knowledge and skill standard which candidates must demonstrate.

Most professional examinations use the Angoff Method to set the passing score. In this method a panel of subject experts individually rate each examination question in terms of what portion of the minimally qualified candidates will know the answer. The ratings among raters and across all questions lead to the passing score. If the examination contains difficult or obscure questions, raters expect fewer people to get those questions correct. If there are easy questions, raters expect more people to get them right. The procedure results in a different passing score for each edition of the examination.

Setting passing score arbitrarily, as is done in determining the score required to pass a college course, is not defensible. Similarly, using normative scoring, such as is done on achievement tests, is not defensible or fair. Other test takers affect who achieves a certain score.

1.4.6 Recertification

Another important criterion for national accreditation is having a process for recertification. The recertification cycle is normally five years or less. One method for achieving recertification is retaking and passing the certification examination. Many certification programs emphasize continuing education and learning in various forms and may allow credit from a menu of professional development activities.

1.5 SUMMARY

Certification is a voluntary process in the United States which helps define competency in various disciplines, such as ergonomics, for which there may be no licensing by states. Certification has grown rapidly as a means to help assure competency in professional and sub-professional practice in many fields including ergonomics. It adds value to practitioners, the profession, employers, government agencies and the public. It can have significant impacts on pay.

Certification typically evaluates education, experience and knowledge and skills essential in practice. When the procedures for evaluating candidate qualifications, knowledge and skills are thorough, they can assure a minimum level of competency. Certification will never predict perfect performance by someone.

National standards for certifying bodies, third party evaluation, and accreditation provide a means to assure those in practice, employers, government agencies and the public that a certification has quality. The accreditation process helps ensure the validity and reliability of practice and ensures that the organization awarding the certification operates with sound business, financial, and governance practices and provides a fair process for all candidates.

The bottom line is that ergonomics certification is a means which helps ensure that people claiming to be in ergonomics practice know what they are doing.

1.6 SELECTED ERGONOMIC CERTIFICATIONS

Below is basic information on selected ergonomics titles.

1.6.1 Certified Professional Ergonomist (CPE)

Date of Origin: 1992
Number Certified: 785
Currently Certified: 760
Accreditation: Working toward NCCA and CESB accreditation
Organization: Board of Certification in Professional Ergonomics
Address: PO Box 2811
Bellingham, WA 98227-2811
Phone: 360-671-7601
Fax: 360-671-7681
Web site:
General Qualifications: ● Master' degree in ergonomics, human factors or
(Contact BCPE for equivalent education
additional details) ● 4 years of professional ergonomics practice
● Single, one-day examination
Recertification: Program under development
Interim Title: The Associate Ergonomics/Human Factors Professional
(AEP/AHFP) is awarded to individuals meeting education
requirement, passing Part I of the CPE examination and
still working to meet experience requirement.

1.6.2 Certified Human Factors Professional (CHFP)

Comment: This designation is an alternate to the CPE designation.
After meeting all requirements for the CPE, certificants
may choose which title they wish to use and to have on
their certificate.

1.6.3 Certified Ergonomics Associate (CEA)

Date of Origin: 1998
Number Certified: 23
Currently Certified: 23
Accreditation: Working toward NCCA & CESB accreditation
Organization: Board of Certification in Professional Ergonomics
Address: PO Box 2811
Bellingham, WA 98277-2811
Phone: 360-671-7601
Fax: 360-671-7681
Web Site: www.bcpe.org
Qualifications: ● Bachelor's degree from an accredited university
● At least 200 contact hours of ergonomics training
● Two years of full-time practice in ergonomics
● Passing score on CEA examination
Recertification: Program under development

1.6.4 Certified Safety Professional (CSP) with a Specialty in Ergonomics

Date of Origin:	1999
Number Certified:	25
Currently Certified:	25
Accreditation:	NCCA and CESB
Organization:	Board of Certified Safety Professionals
Address:	208 Burwash Avenue
	Savoy, IL 61874
Phone:	217-359-9263
Fax:	217-359-0055
Web Site:	www.bcsp.org
Qualifications:	Holding the Certified Safety Professional (CSP) designation. (Note: The ergonomics specialty is not a separate certification from the CSP.)
Recertification:	Required as part of the CSP.
Interim Title:	Individuals meeting the education requirement and passing the Safety Fundamentals Examination are awarded the Associate Safety Professional (ASP) designation to show progress toward the CSP.

HEALTH
MANAGEMENT

CHAPTER THIRTY

Job Simulation and Physical Abilities Testing

Thomas R. Valentine, Manager, Personnel Systems and
John Albarino, Senior Physical Abilities Specialist

Metro-North Railroad

COMPANY INFORMATION

Metro-North Railroad is a subsidiary agency of the Metropolitan Transportation Authority of New York which provides commuter rail service to 100,000 customers per day from New York and Connecticut. Grand Central Terminal in Manhattan is the terminus of the railroad which branches out 75 miles to the east to New Haven CT and 75 miles to the north over 2 divisions to Poughkeepsie and Dover Plains, NY. We operate over and maintain 500+ miles of track utilizing 920 cars and locomotives to transport our customers. The equipment is maintained at 7 major repair shops and outdoor yards and facilities about the property. This requires a workforce of 4,440 people both union and management encompassing 240 job titles with 19 unions represented on the property.

Recruitment for both union and management positions is performed by the employment group and test development and administration is performed by the Testing and Validation group.

BACKGROUND

In 1995, the Training and Operating departments began expressing concern that some new hires and employees were able to pass the standard medical examination for a position but were not able to perform some of the physical functions of their job. Some of the problem areas identified were climbing on and off railroad equipment, lack of strength to lift materials for their job and lack of stamina for prolonged effort such as an entire shift cleaning and mopping railroad passenger cars. Meetings were held with the Training Department, Injury claims Group and the user departments of the Operating Division to discuss their problems. At these meetings, the group decided that an acceptable solution would be to test employees for physical abilities before hire or return to work under job simulated controlled conditions. In other words, test the body using job simulation as we do the mind with written testing.

Medical Guidelines developed in 1992 established standards for cardiac function, lung capacity, vision, strength and other abilities and provided critical and frequent tasks for many skilled craft jobs. The Medical Guidelines project was completed when tasks for all relevant positions were rated on criticality and frequency by Subject Matter Experts. Comparison of the tasks revealed a number of similar frequent tasks and abilities across many jobs including walking along loose surfaces, climbing on and off equipment, hand dexterity and various levels of strength.

WORKSTATION DEVELOPMENT

Members of the Testing and Validation group reviewed the tasks and decided that 5 initial workstations would provide the ability to test for these common tasks and abilities for the jobs with the highest volume of employment activity. The 5 workstations would include: an uneven walk 20 feet long and 2 feet wide with raised blocks 3inches in height on alternating sides that candidates would walk along; a door fitted with a standard railcar closer to simulate passing through the cars; 2 passenger seats and luggage rack from a railcar mounted along a wall; a

stationary railcar window used by train crews; and a 52 inch platform used to simulate climbing on and off railcars and trucks (see Figure 1).

Figure 1 – Multi-Purpose Platform

This platform would be constructed using modular steel framing components so that additional testing equipment could be added as needed. The workstations were designed by Testing and Validation and constructed by railroad workers from the Maintenance of Equipment and Maintenance of Way Departments using mostly material that was on hand including equipment from the cars. Almost every department in the Operating Division contributed to the effort. Design work began in September, 1997 and the facility was completed and ready for the first physical ability testing in April, 1998.

Jobs that had a high volume of new hires each year were selected for the initial physical ability testing. These included Assistant Conductor, Electrician, Car Cleaner, Custodian, and Trackworker. Using the previously collected data on frequent and critical tasks, exercises were developed to simulate these tasks.

These ranged from all candidates walking the uneven walk to simulate the loose ballast along the tracks and climbing up and down the steps on the platform for equipment boarding and exiting to assembling a rail joint section to simulate that trackworker task (see Figure 2).

Figure 2 – Rail Joint Assembly

The duration of the testing and the number of repetitions of each task were determined from the job analysis and Medical standards data previously collected. Each test was reviewed by department supervisors and incumbents to ensure its appropriateness to the job before being implemented. This testing is considered an ability test and is conducted before a final offer and medical examination is performed to comply with the provisions of the Americans with Disabilities Act. Additional tests have been added and currently physical ability testing is performed for 12 positions that have high volume turnover in the company. Candidates for testing are provided an information sheet detailing the tasks to be performed and are required to consent to a release and provide medical approval from their personal physician before testing. The testing room is physically located within the Medical Department so that medical personnel are on-hand in case of an emergency.

CONCLUSION

The testing has been underway for one year and the results have been favorable. The Training and Operating Departments have been very satisfied with the ability of the new hires to perform the physical aspects of their positions. To date, there have not been a significant number of people tested to assess injury reduction and job performance.

CHAPTER THIRTY ONE

Nerve Conduction Testing: Post-Offer Job Assessments

John E. Johnson, Ph.D., CPE

Recently, several methods have been promoted as reliable, objective and inexpensive in testing for signs of CTS in the workplace. Measurements of distal latency are traditionally made by stimulating a peripheral nerve with an electrical impulse applied to the skin over the course of the nerve in question, and measuring the conduction time using a second set of skin electrodes placed over a muscle innervated by the nerve and distal to the point of stimulation. Conduction of the stimulus along the nerve to the muscle causes muscle contraction for the motor fiber testing and the resulting muscle action potential is detected by the skin electrodes, amplified and the signal displayed on an oscilloscope screen. Measurement of the distance between the stimulus and the onset of the action potential on the oscilloscope screen allow calculation of the nerve conduction time, or latency. A portable electroneurometer, such as the NervePace-100, measures the distal latency of the motor nerve across the carpal tunnel and displays the value in milliseconds on a LED readout. Despite considerable interest in these systems, little formal research has been conducted to determine whether they can reliably detect physiologic signs associated with CTS (Grant, Congleton, Kuppa, Lessard, and Hutchingson, 1992). Although standard nerve conduction studies are the most sensitive objective test detecting peripheral compression neuropathy in CTS, they are reported to have a false negative rate ranging between 5 and 20% (Melvin, Schuchmann, and Lanese, 1973). In comparison to more formal nerve studies, the electroneurometer gives no information about the sensory conduction of the median nerve, so that the question remains whether the electroneurometer would adequately reflect the condition of the nerve. These considerations are complicated by the findings that sensory changes often proceed motor abnormalities in CTS and that as many as 20% of cases with operatively proven CTS may have no motor abnormalities (Chrisanne, 1987 and Liebhuber, 1986).

It has always been assumed that the repetitive work was the only cause of carpal tunnel syndrome but current data does not conclusively prove whether or not the job itself is the cause of carpal tunnel syndrome or is only a contributing or fringe element of the malady (Johnson, 1994). Merely echoing an interested party's allegation for or against work-relatedness is unacceptable. A self-reported history suggesting work-relatedness may suggest that a condition is work-related, but does not establish association, causation, or aggravation to a reasonable degree of medical probability. An association between symptoms and work does not confirm or refute a hypothesis of causation between the condition and work (Moore, 1991). Risk for carpal tunnel syndrome is closely related to general physical condition. Individual factors, such as body mass index, age, wrist dimensions, and physical activity, are far more important for determining who develops clinical CTS than are job-related factors, such as specific job, force of repetitions, duration of employment, or industry (Nathan and Keniston, 1993).

BACKGROUND FOR STUDY

Ergonomic handbooks used by industry often list repetitive activities, wrist flexion and high pinch force levels as being risk factors for workers to avoid. Ergonomic hazards relative to work-related musculoskeletal disorder refer to physical stressors and workplace conditions that pose a risk of injury or illness to the musculoskeletal system of the worker. Although epidemiological evidence suggests an association between wrist position and pinch grip with CTS, a direct causal relationship has not been well established (Rosecrance, 1993). Disorders such as CTS can be caused or aggravated by non-occupational factors (Franklin et al. 1991). According to the biomechanical model, the mechanism that may be responsible for the development and precipitation of CTS is compression on the median nerve and the flexor tendons and transverse carpal tunnel (Armstrong and Chaffin, 1979). As industries have become more pro-active, some have implemented objective tests to detect the distal latency of workers prior to entering the workforce at that specific plant. By identifying new hires with slow nerves, prior to starting a job, employment may be temporarily withheld or tasks modified to reduce the effects that repetitive work may have on the median nerve. Nerve conduction studies (NCS) are considered the best objective test for supporting the clinical diagnosis of CTS (Bleecker and Agnew, 1987: Rempel et al, 1992).

OBJECTIVE

The objective of this study was to randomly sample the new hires' population to determine what portion of them are predisposed to CTS, denoted by a slow median nerve and to determine the strength of correlation between electroneurometer testing of the median nerve with the NervePace-100 and formal NCS both of the motor sensory fibers of the nerve. Previous research suggests that sensory studies are more sensitive than motor studies and, therefore, would show slowing prior to motor nerve testing.

SIGNIFICANCE OF STUDY

As the costs of CTS in industry increases, there is a need to accurately detect at the earliest stages which individuals have these disorders. It is important that the testing method used to test slowing of the median nerve be non-invasive, cost-effective and accurate. Accuracy of test methods needs to be quantified to insure that both the employer and employee have their legal rights preserved. With the passage of the

Americans with Disabilities Act (ADA), both employers and employees need to be aware of the ramifications of pre-placement, post-offer testing in order to operate within the guidelines of the law. Here, non-quantifiable injuries or illnesses, i.e., those where no objective medical evidence exists, will be the most troublesome form of abuse, as the medical professional will be forced to rely solely on the claimant's communications to determine the existence of a disability (Barnhard, 1992). According to the EEOC, "[a]n employer may require as a qualification standard that an individual not pose a 'direct threat' to the health or safety of the individual or others, if this standard is applied to all applicants for a particular job." (EEOC manual, supra note 33, 4.5). In general, the employer must be prepared to identify: a significant risk of substantial harm; a specific risk, not one that is speculative or remote; an assessment of risk based on objective medical or other factual evidence regarding a particular individual; and whether the risk can be eliminated or reduced below the level of a 'direct threat' by reasonable accommodation even if a genuine significant risk of substantial harm exists (EEOC manual, supra note 33, 4.5).

STUDY

In setting up the study, the policy of the company has to be adhered to in order to be consistent with all applicants. The NervePace-100 was used as a screening device prior to testing the device against the formal nerve conduction studies for both the motor and sensory fibers. Therefore, the testing was performed during the applicant's pre-placement, post-offer physical. When an employee has been offered a position, the first step for any company or employer should be to conduct a medical exam. The purpose of the physical exam is to ensure that the individual is healthy and able to perform the duties or tasks that may be assigned. A health history can and should be administered by a qualified medical professional. Normally in a plant situation, a nurse has the responsibility of conducting the post-offer physical assessment.

The first part of the physical includes a musculoskeletal exam that has an inspection for redness, swelling, atrophy, nodules, deformities, surgical scars; palpation for heat, tenderness, crepitus, nodules; range of motion for extension, flexion, rotation, and lateral bending. Additional tests that are administered are Finkelstein, Tinel, and Phalen. Although these 3 tests can detect problems or identify a situation, these tests are subjective and depend on the cooperation of the applicant. One situation that spurred the implementation of nerve conduction testing was an employee who was diagnosed and treated for bilateral CTS after working in the meat plant for only one week. This case is not unique and is one problem that companies are facing in the real world. How does an employer protect the post-offer candidate and what is the employers' moral obligation? A giant step is implementing formal nerve conduction testing for post-offer job assessments.

RESULTS

The comparison between the testing methods is highly important. The formal nerve conduction study for sensory conduction detected 36 positive hands from 23 subjects. The formal nerve conduction study for motor conduction detected 19 positive hands from 12 subjects. Testing by the NervePace-100 detected 28 positive hands from 20 subjects. Comparison between the tests shows that there was very poor correlation between the NervePace-100 and the formal nerve conduction testing.

DISCUSSION

As the results have shown, there are some issues to address regarding the use of nerve conduction testing and the types of testing that are available. This research in a plant environment has reflected that there are a number of applicants that have slow distal latency of their median nerves prior to working in a meat-processing environment. Secondly, there is a poor correlation between the portable electroneurometer and formal median nerve conduction testing. Therefore, a good possibility presents itself for false negative readings and false positive readings when conducting post-offer nerve conduction testing. The problem arises when the false negative readings do not detect the possibility of developing CTS and prevents employers and employees to take the proper action.

Formal nerve conduction testing can be a highly beneficial tool for employers and employees. These tests can detect abnormal nerve condition to make all parties aware of the possible risks that may be involved in being placed in certain tasks. The ADA is not violated or compromised in this case. The purpose of the ADA is to protect workers with disabilities and to prevent discrimination against them. By implementing nerve conduction testing as part of a post-offer physical assessment for all workers, there is no basis for the argument that applicants are being discriminated against. As more work and data is collected in the industrial workplace, greater knowledge can be gained on the comparison between work, individual factors and CTS.

REFERENCES

Armstrong TG, Chaffin DB. Some biomechanical aspects of the carpal tunnel. *J Biomechanics.* 1979; **12**:567-570.

Barnhard TH. Disabling Americans: Costing out the Americans with disabilities act. *Symposium: Enabling the workplace: Will the Americans with disabilities act meet the challenge?*

Bleeker ML, Agnew J. New techniques for the diagnosis of carpal tunnel syndrome. *Scand J. Work Environ Health.* 1987; **13**:385-388.

Chrisanne G. et al. Electrodiagnostic characteristics of acute carpal tunnel syndrome. *Arch Phys Med Reha.* 1987; **68**:545.

Federal Register. Department of Labor: Occupational Safety and Health Administration. 29 CFR Part 1910: *Ergonomic Safety and Health Management; Proposed Rule. 1992*; **57**:34192-34200.

Franklin GM, Haug J, Heyer N, Checkoway H, Peck N. Occupational carpal tunnel syndrome in Washington State, 1984-1988. *AJPH.* 1991; **81(6)**:741-746.

Grant KA, Congleton JJ, Kuppa RJ, Lessard CS, Huchingson RD. Use of motor nerve conduction testing and vibration sensitivity testing as screening tools for carpal tunnel syndrome in industry. *J Hand Surg.* 1992; **17A**:71-76.

Johnson J. *A comparison between formal median nerve conduction studies and a portable electroneurometer.* Unpublished doctoral dissertation. LaSalle University. 1994.

Melvin JL, Schuchmann JA, Lanese RR. Diagnostic specificity of motor sensory nerve conduction variables in the carpal tunnel syndrome. *Arch Phys Med Rehabil.* 1973; **54**:69-74.

Moore SJ. *Clinical determination of work-relatedness in carpal tunnel syndrome.* Unpublished. 1991.

Nathan PA, Keniston RC. *Carpal tunnel syndrome and its relation to general physical condition.* Hand Clinics. 1993.

Rempel DM, Harrison RJ, Barnhart S. Work-related cumulative trauma disorders of the upper extremity. *JAMA* 1992; **267**:838-842.

Rosecrance JC. *The use of provocation for the detection of abnormal nerve conduction in carpal tunnel syndrome.* Unpublished doctoral dissertation, University of Iowa. 1993.

DESIGN

CHAPTER THIRTY TWO

An Ergonomic Approach to Facility Design and Layout

Bill W. Brown

1.1 INTRODUCTION

Facility design and layout is an area that could be greatly benefited by Human Factors and Ergonomics. All too often the facility is designed without thinking past the general use of the space. The following case study will demonstrate the impact of approaching facility design from an ergonomist's point-of-view.

1.1.1 Background

A doctor's clinic for women was expanding their current facility from 7,000 to 12,000 square feet. The clinic had already been working with some architects to establish a floor plan and design concept. Neither the architect nor the clinic staff was comfortable with the conclusions they were making in the construction plan. They also did not know how to measure the effectiveness of the design with respect to workflow, productivity, or comfort of the staff and patients.

As the Human Factors professional on the team, I joined the effort after an initial plan was created and associated problems existed. The architects and clinic staff did understand that it would be easier to invest a little more effort up front to create a solution they were more satisfied with than to proceed and end up with something they knew would not work.

1.1.2 Objective

The purpose of involving a Human Factors engineer on the design team included applying principles of engineering design and psychology to:
- Understand the needs of the staff who will be using the space,
- Understand the needs of the patients/customers who visit the facility,
- Understand the goals of the design committee,
- Work with the design committee to identify alternative design concepts,
- Create a workspace that helps the staff accomplish their work,
- Identify potential problem areas in the design of the new facility, and
- Evaluate the design to determine how well the goals were met.

1.2 METHODOLOGY

The design team was made up of the architects, doctors and administrative staff at the clinic, and the Human Factors professional. The design team was assembled to resolve already identified problem areas and other potential problem areas as the project progressed. The Human Factors support consisted of data collection, conceptual design, and analysis.

1.2.1 Data collection

Probably the most important phase of the project was finding out about the work that had to be performed in the facility. We talked with the people who would be using the facility. Since the people were already performing their duties in the current facility, we were able to observe them in the environment. Our data collection was targeted at understanding the priorities and values of the people (e.g., doctors, nurses, and patients) who would use the space, the work processes used in the different functional areas, and the constraints. In trying to find out what the staff did, we had to ask why they did certain activities. We had to distinguish between what was being done for some reason and what was done simply because it had always been done that way. Data collection activities included a walk-through of the facility, meeting with the architects, meeting with the design committee to discuss the goals, and staff interviews.

1.2.1.1 Facility walk-through

This included walking through the current facility and the new leased space. On the walk-through of the current facility, we learned how the work is currently performed. We tried to learn as much as we could about what worked well and what did not. The new facility was initially only an empty, open space with pillars distributed across the floor to support the floor above. An elevator shaft existed in the center. A fire escape was positioned along the back wall. Some of the windows had more desirable views than others. Visiting the new leased space identified some of the structural requirements and constraints we needed to understand before making design recommendations.

1.2.1.2 Meeting with the architect

We met with the architects to understand what was flexible and the areas that were physically constrained. We needed to understand the goals the architects had identified and established so that our recommendations were not in conflict with values already assigned to the project. We also needed to understand what form of communication would be most useful for the architects to incorporate our recommendations into the final construction plan.

1.2.1.3 Meetings with the design committee

The design committee was made up of people who understood the work of the facility and the goals of the project. During the meetings, we discussed the current goals, the problem areas identified, and presented some design concepts to determine how well they met the needs of the people who would be using the space. We worked with the design team to evaluate the design concepts and create boundaries that helped move the design toward a final construction plan. The design team was challenged with trying to represent and resolve the pros and cons of each of the functional areas as trade-offs were made and still move the project forward.

1.2.1.4 Staff interviews

We met with the clinic staff that represented each of the functions performed in the doctor's office. The staff met with us individually, and as a group, to discuss the way they did their work and their expectations for a new facility. We used a task-based approach with the staff individually and asked them to talk through the steps of the task they had to perform Discussions were targeted at getting the staff to think out of the box. What physical requirements were necessary to help them complete their job? Where did equipment and supplies need to be positioned to facilitate productivity? How did they need to interact with other functional staff? What part of the task had the potential for making errors?

1.2.2 Conceptual design

Conceptual design was used to help the design team step back from the details to get a solution to the bigger problem. We used a modification of a method used in Japan called Kan-sei engineering to facilitate this process.

Kan-sei engineering is the Japanese term used to describe a suite of specialized principles, techniques and processes that has helped lead Japanese technology design and production to world prominence. A Kan-sei designed product is distinguished by its deliberate capability to respond to a characteristic attitude or 'set' of the user that predisposes him or her to appreciate things in a certain way. Kan-sei design is responsible for some of the most successful technical products Japan now markets to the U.S. and Western Europe.

Our process started with simple shapes that might afford the functionality needs. We evaluated the alternatives to establish the pros and cons of each concept. We tried combining shapes to capture the characteristic advantages of each to see if we could obtain a better overall result. We then started inserting more detail to make sure the results would work for our intended purposes without distorting size or negatively affecting spatial relationships.

1.2.3 Human factors and ergonomic analysis

The data collection activities helped identify the type of tasks (or user goals) the facility needed to accommodate. In our analysis of the facility, we were trying to make sure the activities identified could be accomplished and that the solutions would help meet the established goals and values.

1.2.3.1 User centered analysis

Every individual at the clinic had goals they must accomplish to complete their job and positively contribute to the overall process. Our analysis centered on identifying their job responsibilities and the ultimate goal(s). The key to this analysis was identifying the ultimate goal of the individual and possible alternatives that would allow them to accomplish the subtasks along the way. As we documented the goals and steps required to accomplish the goal, we looked for:
- Information needed to know what to do at each step,
- Physical requirements (i.e., spatial location, color, size, shape, etc),
- Sequences that had to occur in a specific order,
- Safety issues,
- Steps that could be performed improperly,
- Feedback required so the individual knew the step was complete (or not).

1.2.3.2 Workstation analysis

Many of the functional areas were similar but columns and other support structures in the space created some unique situations. We analyzed the workstations to make sure they fit the current workers. We also tried to set up solutions that would fit new workers as needs change or new employees join the team. Because of a tight budget, we focused on workstations that were similar throughout the facility and the ones that posed the highest risk (e.g., workstations that are used the most). An evaluation of all workstations would have been beneficial.

1.2.3.3 Workflow analysis

The workflow evaluation included physical exchanges of paper and people and communication flows between staff and patients. We analyzed the current workflow and made sure the new facility would work for the current process. In addition, we made suggestions intended to improve the overall workflow. Since more than one individual was involved, we looked at transition alternatives.

1.2.3.4 Construction plan analysis

We compared our list of tasks that must be accomplished to the current construction plan to make sure that it would allow the staff to accomplish their goals. As the plan changed over time, we provided design suggestions for problem areas and reviewed the plan to make sure new issues were not introduced.

1.3 RESULTS

We talked with individuals in each of the functional areas to understand the workflow across the clinic and within the specific workspace. Results were shared and discussed with the design committee to determine the impact on the construction plan. The design committee used the results to brainstorm possible solutions and resolve trade-offs as conflicts were identified.

1.3.1 User goals

The user-centered analysis identified a number of goals the users of the facility might have. Users include doctors, nurses, administrative staff, and patients. Some of the goals involved only one user while others describe interactions between multiple users. Some examples of the goals identified include the following:

Administrative staff goal: Locate patient chart
Patient goal: Get to clinic
Patient goal, involves administrative staff: Check-in
Nurse goal, involves patient: Lead patient to destination
Doctor goal, involves patient and nurse: Perform check-up
Nurse goal, involves patient: Give injection
Technician goal, involves patient: Perform ultra sound
Patient goal, involves administrative staff: Checkout
Patient goal, involves administrative staff: Schedule an appointment
Goal initiated by patient or administrative staff: Resolve bookkeeping question

By understanding the steps required to accomplish the goal, we identified some design constraints that influenced spatial relationships of functional areas and equipment. While it may be useful to the reader to see all the results, we will only use "Check-in" as an example of the process. The steps for checking in are::
Step 1. Patient arrives
Step 2. Receptionist locates patient chart
Step 3. Patient updates information if necessary
Step 4. Receptionist copies medical card or authorizes Medic-Aid card
Step 5. Receptionist identifies purpose of visit (i.e., shots, exam, billing, etc.)
Step 6. Receptionist notes time of arrival and any info given by patient on chart
Step 7. Receptionist puts chart in appropriate destination for routing
Step 8. Receptionist notifies appropriate parties (i.e., bookkeeping, nurse, etc.)

Listing the needs of the user and the potential problems at each step helped identify design requirements for the facility. To illustrate the kind of info that was useful the following sections provide some highlights using the example goal of checking in.

For step 1 (patient arrives) the patient needs to know where to go to initiate the purpose of the visit. The receptionist needs to be able to recognize that a new person has entered the facility and be able to make himself or herself available to the individual. Before proceeding the new arrival must provide their name and purpose of the visit to the receptionist. Some patients will provide this information without a request. Other patients will need prompting for the required information.

For step 2 (receptionist locates patient chart) the staff had already established a process. The charts for all prescheduled appointments were in one location sorted by doctor and time of appointment. If the appointment had been made in the last hour or the patient was a walk-in, the receptionist may need to request the chart from records. Since a request of this type was common in a normal day, a top priority was to locate medical records fairly close by to the reception area.

For step 3 (patient updates information) the clinic needed the most recent information. In particular, the clinic needed to verify that the patient's address and insurance provider were correct. The patient was asked to verify the information, particularly if it had been more than three months since the last visit. Since the information was hand-written during the first visit and any updates were penciled in, there was always a chance the handwriting could not be interpretable when entered into the computer system. However, the potential for errors was not so frequent or severe that it justified a more automated system.

Step 4 (Receptionist copies medical card or authorizes Medic-Aid card) required a copy machine and a Medicaid authorization machine in the vicinity. A copy of the card was only needed if it was not already in the chart. Since it is easier to respond to a cue to do something than it is to recognize that something is missing, it would be better to include a note in the charts that needed a copy of the card or Medicaid authorization when the chart was assembled. The cue could be removed when the copy was placed in the chart.

Step 5 (receptionist identifies purpose of visit) requires the receptionist to determine the purpose of the visit. Even though the patient may have stated a purpose, there may be other tasks that must be accomplished during the visit that may or may not have been known by the patient. The receptionist must assess the entire purpose and initiate each task in the most optimum sequence. For example, if the patient is there for an exam and bookkeeping needs to visit with them as well, the receptionist needs to determine if the patient has time to get in to bookkeeping before the exam or if they should wait until afterwards.

Step 6 Receptionist notes time and any info given) requires the receptionist to enter the time of arrival. Some procedures must be done within certain time limits. An official time must be used that all clinic staff rely on. This may require a clock visible in the reception area to receptionist, patient, and nurse.

Some patients believe that what they told the receptionist was passed on to the nurse and doctor. Some patients are put out when they have to repeat their story. To provide a prompt for the nurse or doctor, any information stated at check-in should be provided in some way in the chart. The note may not become an official part of the chart, but it would notify the nurse or doctor of the situation.

Step 7 (Receptionist puts chart in appropriate destination for routing) requires a specific location for each chart. The chart might need to be passed to a nurse, bookkeeping, ultra sound technician, etc. If the chart is placed in the wrong location, it may cause a delay in the process.

Step 8 (Receptionist notifies appropriate parties) requires the receptionist to notify all parties that might be involved and coordinate the entire visit. Many patients want an estimated waiting time. The receptionist must be able to estimate the waiting time and how long before the patient is finished in order to inform the patient. Tracking the whole process as it transpires can be complicated if multiple parties are involved during the visit.

1.3.2 Overall workflow

The flow of people and paper through the facility was a large contributor in determining the appropriate layout of the new space. The workflow models were developed from the perspective of each person involved in the process for both physical relationships and communications. The workflow diagrams helped us talk through how and when records and patients would be handed off.

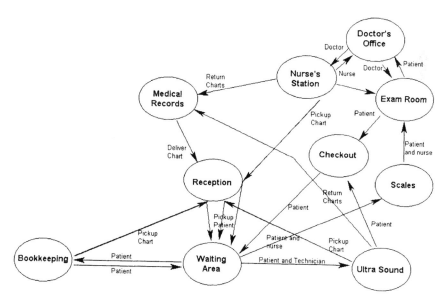

Figure 1 Physical relationship model

The physical relationships are shown in Figure 1. The model was developed by looking for the areas each functional area needed to physically interact with. The bubbles representing the functional areas were then moved around to minimize the distance and amount of overlap.

The communication relationships are shown in Figure 2. The model was developed by looking at what functional areas needed to communicate for some reason. If the communication did not include visual communication, then the physical model was not changed. The functional areas with solid lines indicate the areas that might need to be located close to each other for visual communication.

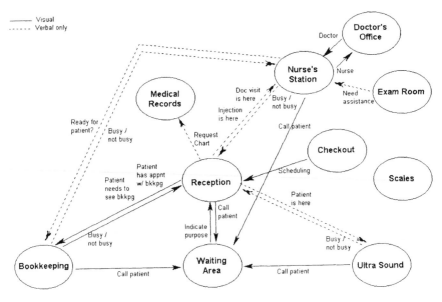

Figure 2. Communication model

1.3.3 Communication with design committee

Previous experience had demonstrated that architects and engineers learn to read overhead floor plans because of the nature of their work. Even some of these professionals have difficulty interpreting floor plans on occasion. To the design committee's credit, many of them learned to read the architectural plans. However, throughout the design process, we found it useful to prepare three-dimensional perspective drawings of the workspace for the design committee. The perspective drawings made it easier for the design committee to understand and interpret the concepts and specific details. An example of these perspective drawings is shown in Figure 3.

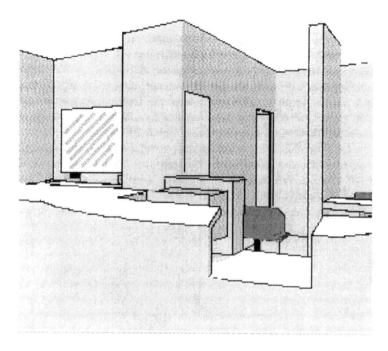

Figure 3 Example of drawings provided to design committee

1.3.4 Communication with architects

As a part of the evaluation, we found that including transparent overlays of the specific design concepts printed at the same scale as the architectural drawings helped the architect to capture specific details in the construction plan. The transparencies could be positioned over the architectural plans and used to quickly identify missing details, sketch new concepts, and assist in standardizing similar spaces. Spatial drawings also included a specification sheet for details within the space similar to what the architects were used to seeing.

1.4 CONCLUSIONS AND RECOMMENDATIONS

Our design recommendations were based on our understanding of the goals of the design committee, the needs of the staff, and expectations of patients. To resolve tradeoffs in design solutions between functional areas, we referred back to the user centered analysis and workflow diagrams. The following sections represent some of our conclusions.

1.4.1 Conceptual design

A lot of the problem areas centered on the reception area. This one area would be the center of the workflow and interaction between different people in the clinic. It is the entrance to the facility and provides the first interaction for the patient with the facility. It had to invite the patient to enter, provide clearly defined areas for check-in and checkout, and eliminate, or at least minimize, congestion. The reception area is where the paper trail begins for a patient's visit. The receptionist must interact with medical records, nurses, and bookkeeping.

The reception area had so much focus that it was easy to get bogged down in the details. We needed to be able to step back from the detail to view the big picture. Kan-sei engineering techniques were used to draw out possible alternatives that would meet the needs of the facility. Our objective was to create a solution that allowed each of the functions listed above. Some of the concepts proposed started more as a shape that allowed for interactions on multiple sides (see Figure 4). Our final design combined a couple of concepts to capture all the advantages.

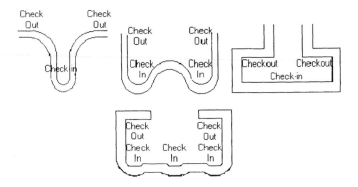

Figure 4 Reception area concepts

1.4.2 Relationship of functional areas

The physical relationship between functional areas can have a large impact on the way the individuals at the clinic complete the job. There are always trade-offs in the final design, but one location might be optimum for one group and negatively effect productivity for another. We wanted the best possible solution for everyone involved. A good example of how we determined the relationship between functional areas was bookkeeping.

Bookkeeping tracks the financial interactions with the patient. They talk to each new patient to establish a payment plan and, as necessary, before or after future visits. Many patients are aware that they need to see bookkeeping when they come in for a visit.

From the physical model of the workflow, the big revelation was that the tradition of putting bookkeeping in the back needed to change. Since bookkeeping needed to interact with reception, medical records, and the patients in the waiting area, we made them accessible from the waiting room and visible to the receptionist. This would allow some patients to step into bookkeeping and take care of their business without talking to the receptionist.

1.4.3 Managing congestion

The clinic had always had one entrance (see Figure 5) from the waiting area into the clinic. Not only was this the only way in and out, but patients standing at checkout often blocked the exit for others.

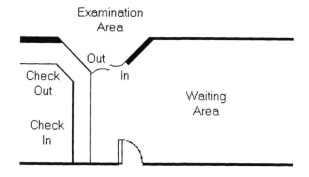

Figure 5 Entrance to clinic in previous facility

The clinic was already dealing with 30 patients every hour checking in and out. Since the new facility was expected to increase the traffic, we had to minimize any potential bottleneck areas. We did not want patients waiting in lines or having to navigate through crowds. For administrative reasons, the clinic also wanted to make sure everyone was able to checkout when they were finished.

The new facility was designed with two passages from the waiting area into the clinic. We considered creating circular pathways through the facility that led back to a single checkout. However, patients were generally allowed to dress and leave when they were ready. Patients were more comfortable retracing the way they entered the facility when they were on their own. We decided to place checkout near each entrance/exit but provide an area for the patient to step out of the main hallway (see Figure 6).

By simply splitting up where the patients entered from the waiting area, the traffic was reduced by half. Another advantage of this concept was that on light days the clinic could simply operate on one side of the facility. If, at some time in the future, the needs changed, then the clinic could be reduced to one side of the building.

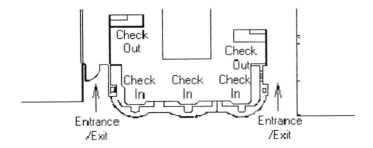

Figure 6 Reception area in the new facility

1.4.4 Reduce confusion by utilizing defined space

With the long counter in the reception area, we wanted to allow patients to interact with the receptionist from any position that was convenient for them. However, we wanted to create a location that would indicate where the patient should stand during check-in so that the receptionist could talk to the patient and see the computer monitor without turning or twisting.

We decided to integrate the following features into the counter. Rather than one long surface, we created three distinct areas for interaction by spacing three circular pods into the counter (see Figure 6). The majority of the counter was built 43" high to create a writing surface for the patients. The writing surface concealed the equipment, particularly cabling, used by the receptionist. The top of the monitor was visible from the waiting area side. An opening just to the side of the monitor at work surface height of the receptionist invites the patients to step to the side of the monitor so that the receptionists are not looking over the monitor (see Figure 7). The two outside workstations are balanced with the same openings and high writing surfaces.

Figure 7 Cutout in counter to invite patient to interact
(proposed drawing compared to actual)

1.4.5 Combine aesthetics and functionality

The clinic needed to centralize copiers, printers, and other equipment in a place so that all receptionists could easily utilize them. The most optimum location for this equipment was also the same location that everyone entering the facility through the reception area would see. Even though people realized that it took all this equipment to run the facility, they did not want to see it all the time. We wanted to make the best first impression on those visiting the facility, and we did not want the equipment to be what they remembered.

Cabinets were designed to conceal the equipment while still making it accessible. Doors on the cabinet disappeared down the interior sides of the cabinet to expose the equipment and not interfere with traffic flow around the corners.

From the front, as viewed from the waiting area, the patients see smooth, symmetrical columns. A counter between the columns provides additional storage and conceals other centralized equipment, which is also easily accessible to the receptionists. Over the counter is a one-way mirror with the company logo embossed in the mirror. The mirror allows phone staff to monitor the reception area without being seen or heard by patients (see Figure 7).

1.4.6 Visual obstruction to break lines of sight

In many areas of the clinic, we needed to give special consideration to privacy. We needed to ensure that someone could not intentionally or unintentionally view confidential information that may be displayed on a computer monitor or see into an exam room at a time when someone might be indisposed.

Because of the orientation of check-in and checkout, patients at either location could potentially view information displayed on a monitor at the other location. We recommended a ceiling height wall to divide the two work areas. The wall minimized the patients' view of the monitor while checking in or checking out. However, it still allowed staff members to communicate and pass paper work.

Exam rooms were intentionally staggered so that someone sitting in one exam room with the door open could not see into an exam room across the hall.

Figure 8 Overhead view of exam room

Each exam room was designed with the exam table behind the door so that the door had to be open more than 90 degrees before the person entering could see anything on the exam table. The room was also designed with a built-in closet. If the patient was still changing when someone opened the door unexpectedly, then the closet door provided some protection from view (see Figures 8 and 9).

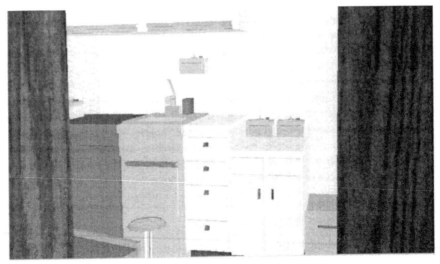

Figure 9 View into exam room with main door and closet open

We also used this principle to give the nurses a chance to review the charts before picking up the patient (see Figure 10).

Figure 10 Columns used to obstruct patient's view

Because of the way the charts were passed from reception to nurse to doctor, the nurse did not see the chart until the patient was in the waiting room for the appointment. The chart was placed in a slot just to the side of the reception area, and the nurse would be notified. The nurse picked up the chart on the way to the waiting area. Since it was common for other waiting patients to try to get a nurse to answer a question while they were in the waiting room, the nurse needed a place to review the chart before stepping into the waiting area. We placed ceiling columns at both ends of the reception desk so that waiting patients could not see nurses getting charts from the slots.

1.4.7 Individual workstations

Workstations throughout the facility varied slightly. However, certain items were recommended for all workstations.

Work surfaces were built into the facility. Since the height of the work surface was fixed, the other surfaces in the work area had to be adjustable (e.g., keyboard platform, seat height, foot rest, and monitor height).

Requirements for chairs were recommended so that the chairs provided adequate support and would fit all individuals that might use the workstation. Since the clinic already had chairs in the previous facility that met most of the requirements, they stayed with what they had and kept the requirements for future purchases.

An adjustable platform was recommended for keyboard and mouse use. The platform needed to be large enough for the keyboard and a mouse (minimum 27" wide, recommend 30"). Height (and angle) should be adjustable, separate from work surface and monitor platform. An articulating arm was installed for most workstations. However, not all nurses and administrative staff wanted to use them.

We pointed out that the monitor support should be adjustable, separate from the keyboard platform and writing surface. However, we indicated that height of the monitor could be adjusted by using supplies generally available in a typical office (e.g., reams of paper, phone book, etc.). The clinic did not provide any special equipment to support the computer monitors at workstations.

Many of the workstations incorporated overhead cabinets for storage. We recommended the bottom of the cabinet be no higher than 50" from ground level. The underside of the cabinet should incorporate task lighting for illuminating the work surface.

A phone was needed at most workstations. We suggested that the phone should be available near each computer terminal. Suggested placement would be on the left side of the computer for right-handed people to minimize the repetitions with the primary hand. However, the specific placement was up to the individuals' preference.

File cabinets were integrated into the workstation as necessary and oriented so that the individual could open the cabinet from a seated position while working as they would when they needed to access the files.

1.4.8 Provide a quiet area for semi-private work interaction

All of the staff in bookkeeping had been placed in a single open room in the previous facility. The new facility still had all the bookkeeping staff in the same room, but partial walls provided a place for private discussions with patients and separated the workstations. Some of the nurses' stations integrated a chair for visitors. Not all nurses had the need to speak to patients at their workstation. Even with the limited space in the exam rooms, we took advantage of what we had. We already had a stool for the doctor and a build in seat similar to a window seat for the patient. A pull out tray was integrated into the countertop for a writing surface.

1.4.9 Aid training time through consistency

The clinic was set up with three exam rooms per doctor and nurse team. Each room was organized the same so that the doctor or nurse would not have to waste time finding supplies or equipment each time they changed rooms. Also, if a nurse was helping out a different doctor in a different set of exam rooms, then the nurse could expect to find supplies and equipment in the same location. The clinic also has two ultra sound rooms arranged identically so that technicians did not have to spend time locating supplies when they changed rooms. Some discussion centered on whether a room should be mirrored or simply identical. Most of the medical industry seems to be handed-conscious. Once a nurse or a doctor is used to having tools or equipment set up on a certain side they tend to favor that side the rest of their career. We decided to make them all the same.

1.4.10 Communication between receptionists and nurses

Receptionists needed to communicate with all the groups, but it was much more common for the receptionists to be contacting a nurse about a patient who was ready for an exam. The nurse was not usually at her workstation, but she would be expecting a visit. Because of the need for other functions to be closer to the reception area, the nurse's station and exam rooms were located furthest from the reception area. The nurse did not want to make unnecessary trips to the reception area. The scheduled visits were so common that the receptionist usually did not need a response from the nurse. They simply needed to notify the nurse that the patient was ready.

To cue the nurse that a patient was in the waiting room without interrupting other activities, we incorporated a lighting system between reception and nurse's station. When the chart was placed in the slot for the nurse, the receptionist turned on the light. When the nurse picked up the chart, the light was turned off. We discussed a more automated system that was linked to sensors in the box, but it seemed too expensive for the intended purpose. Pressing a button turned on the lights. The receptionist had one button for each slot inside the reception area. Another button for each slot was placed outside for the nurse. The buttons on the outside wall were positioned above the slots and covered with an access panel so that curious children could not activate the switch.

The receptionist could see both the light and the chart to make sure that the chart was picked up. If the chart stayed in the slot for an extended period, then the receptionist could follow-up with the nurse by phoning or paging.

The location of the light at the nurse's station needed to be visible to the nurse while sitting at the desk doing normal tasks or walking around between the exam rooms and doctor's office. The nurse needed to be able to tell if the light was on or off regardless of lighting conditions in the room.

1.4.11 Exam rooms

The exam rooms needed to be set up for doctors of different physical size. Some commonality was necessary so that doctors and nurses do not have to change their entire work process when they move between exam rooms.

The exam table was angled away from the corner of the room. Main access to the exam table was from the right and foot of bed. The doctor needed access to tools stored in the foot of the bed.

Accessories that were needed in close proximity to the exam table included task lighting, ophthalmoscope, and blood pressure cuff. All of these needed to be available on the doctor's right side when standing at the foot of the table. They all need to be able to reach any part of the exam table. All of these accessories have cables and/or power, which need to be managed so the doctor or nurse does not have to untangle them each time they use them. The cables should also not interfere with access to the left side of the bed.

Custom cabinets in the exam room satisfied other needs that are common in health care facilities. The counter needed to incorporate a sink with soap and paper towels close by. Other accessories that needed to be accessible on the counter include tissues, rubber gloves, lubricant, and a sharps container. Accessories were mounted on the wall so they were available without just being open and cluttering the counter space. The cabinets under the sink integrated special compartments for dirty linen, garbage, hazardous waste, and dirty tools to remove clutter from the exam room floor.

Dirty linens could be placed in the cabinet through a swinging door hinged at the top on the face of the cabinet. Inside the cabinet is a container that catches all the dirty linen to minimize further contact with the dirty linens. A hard container would fill the space to catch the linen and could be washed out after use and returned. A soft bag would be an alternative but would require additional effort to catch the linen and would need to be laundered after use. Dirty laundry is removed through the front and carried or rolled to a central location for pickup.

A garbage receptacle and hazardous waste bin is built into the same cabinet. Garbage is inserted through a round hole in the top and hazardous waste through a hinged door on the front. A foot pedal in the base of cabinet opens the hinged door so that hands are not necessary. Front of cabinet opens to remove hazardous waste and get to garbage container behind.

A place for dirty tools is built into the front of the counter below the sink. A foot pedal in the base of the cabinet opens the hinged door so that hands are not necessary. Front of cabinet opens to remove dirty tools and maintain the container with a cleaning solution.

The top of the hinged doors might be aligned with other horizontal lines on the front of the cabinet but the door itself should be different from the other doors on the face of the cabinet so that the doctors and nurses do not mistakenly put an item in the wrong cabinet (e.g., dirty tools in with the linen).

1.5 SUMMARY

Our analysis was able to impact many areas across the doctor's clinic including reception, medical records, bookkeeping, exam rooms, ultra sound, and the administrative offices. Normally, we would get involved earlier in the process. However, this project allowed us to perform the necessary analysis to make the appropriate recommendations before proceeding. The flexibility of the design team made it possible to positively influence the project overall.

The maximum number of patients the clinic had seen in one day in the previous facility was 225. The day was full of chaos and noise. When the clinic moved to the new facility, the number of doctors increased by one; however, the second day in the new facility, the staff saw 210 patients. Within the first week, the clinic had a day they saw 260 patients, and the staff felt they were within their comfort zone. Chaos was minimal. Patients were not required to wait in lines. Supplies were located in natural areas that reduced confusion and allowed the staff to handle the situation. The maximum number of patients they will be able to deal with in the new facility will likely be much higher.

Since every facility has different needs, each solution must be different and must involve the people in the design process that will use the facility. This project demonstrated that coming up with the best solution meant involving the people that would use it. Communicating these needs across functional areas helped other staff understand how their work affected others and how they could help each other. If the designers can tap into the ideas of the people that will use the facility and raise the awareness of how what they do affects others in the facility, then a reasonable solution can be met that is attractive to everyone.

1.6 SUGGESTED READINGS

Nagamachi, M., 1991, Image Processing and Kan-sei Engineering. In *Proceedings of the 22nd Conference on Image Science and Engineering,* pp. 238-288.
Schneider, G. and Winters, J. P., 1998, *Applying Use Cases, A Practical Guide,* (Massachusetts; Addison-Wesley)

CHAPTER THIRTY THREE

Design of a Hoist for the Removal and Replacement of Horizontal Semiconductor Wafer Furnace Heater Cores at the Hewlett-Packard, Corvallis, Oregon Site

Daryl R. Meekins, BSME, EIT
Ergonomics Engineering, Hewlett-Packard

ABSTRACT

Hewlett-Packard lifting guidelines and analysis using common ergonomic evaluation tools indicated that manual replacement of heater cores in horizontal furnaces was unacceptable and engineering intervention was required. The design was guided by a number of physical constraints and worker requirements, for example, weight of and access to the heater cores, accessibility to maintenance bays, and the need for elevated work platforms. The cooperative risk reduction effort took a year and involved maintenance technicians, HP's ergonomics engineer, and engineers from the hoist vendor. The final design employs a strap and roller end-effector with a two-arm articulated manipulator mounted to a counterweighted ball screw type vertical hoist. The hoist has the ability to hold the heater core securely while allowing technicians to place it in position for installation. It is also compatible with the requirements of a Class 100 clean room, is powered by an internal power supply with integral charging unit, and has casters which allow for mobility. Using the hoist, the technicians can now perform the replacement procedure without manual lifting of the cores or exceeding standards for lifting and pushing forces. A follow-up survey indicated a high level of satisfaction by the users of the new system.

PURPOSE AND TARGET AUDIENCE

This paper should familiarize the reader with the problem that generated the need for this project, the constraints that guided the design and fine-tuning the process once the prototype was built. Also included are the details of the role of the Oregon-OSHA Worksite Redesign Program, describing how that program was used to assist with the project. The paper is intended to assist the ergonomist in breaking into the process of designing lifting aids.

INTRODUCTION

Maintenance Technicians in the Furnace area of Corvallis HP integrated circuit factory recognized the undesirable nature of the task of replacing heater cores in horizontal furnaces long before this project began. They had been unsuccessful in finding a commercially available solution to the problem. In June of 1998, a maintenance technician at the Hewlett-Packard Corvallis plant visited on-site medical personnel complaining of shoulder pain. Questions by the nurse as to the maximum weights lifted in the course of typical maintenance tasks revealed a task that the nurse recognized to be far outside lifting standards established by the Environmental Health and Safety (EHS) department at the plant. A subsequent meeting between the nurse, the technician management staff, and engineers determined that the task of manually replacing the heater cores in the horizontal furnaces was too dangerous to be continued, and therefore, the task was "suspended" until intervention was in place. Given the potential impact on production as a result of not performing this maintenance task, an immediate design effort was initiated.

ERGONOMIC ASSESSMENT

The heater cores in the horizontal furnaces weigh 210 pounds each (95.25 kg), are 71.5 inches (181.6 cm) long and 16 inches (40.6 cm) in diameter. They are arranged in a vertical stack of four furnaces with the height of the core centers ranging from 32 inches (81.3 cm) to 86 inches (218.4 cm) above the floor. On average, one to two tube replacements occur per month. In this case, frequency of the lift was not a factor, in the sense commonly used in typical ergonomic evaluations.

Prior to the start of this project, average height and shorter workers were required to use scaffolding intended for maintenance of other tools to support one leg while standing on the frame of the tool with the other (see Photo 1). Technicians removing the core with such precarious footing were exposed to the hazards of being crushed by a falling heater core and falling into sharp sheet metal protective shrouding or possibly contacting electrical lines. They also risked injury from lifting such a heavy, large and awkwardly shaped object over a large vertical and horizontal range of motion. The typical (informal) manner of removing and replacing the two highest cores was to get the tallest technicians in the fabrication facility (hereafter referred to as the "fab"), who could lift the cores out without use of scaffolding. Shortly before the reported injury referred to above, the tall technicians had transferred to another site.

EVALUATION

Hewlett-Packard Company guidelines at the Corvallis facility limit the weight that a technician can lift to 50 pounds under ideal conditions, defined as low frequency (less often than once per hour), close to the body, between elbow and shoulder

Photo 1
Beginning core disassembly

height. Considering these guidelines, even if the heater cores could be held in an "ideal" position with good handholds throughout the replacement procedure (which it cannot, as is evident from Photo 1), and the load could be equally shared, it would still require five technicians to perform the task. Ergonomic and safety considerations including footing, lifting postures, distances to move the core, etc., caused estimates of effective safe team size for the task to grow to between 8 and 10 technicians. Logistically, this was not possible, even if there had been suitable workarounds for these issues.

Two ergonomic evaluation tools were used to estimate the risk of the task. Assumptions for the analysis were that of a 180 pound male (the average team member weight and gender) lifting ½ the total weight of the core, as would nominally be the case in a two-person lift. A NIOSH revised lifting equation analysis of the task resulted in a recommended weight limit (RWL) at the initial position (the core in the furnace) of 10.5 lb. and at the destination (on a common cart in the fab) of 17.8 lb. The lifting index (LI) at the origin was, therefore, about

10 (that is, 10 times the recommended weight limit for the task) and near 6 at the destination, obviously far from an appropriate lift. Use of the University of Utah Simplified Shoulder Moment Estimation tool calculated the maximum available shoulder moment for a 183-lb. male and the postures necessary for this task at 743 in-lbf. (inch-pounds-force, a unit for torque or moment). The actual shoulder moment (worst case) generated by this task was 1420 in-lbf. resulting in a percent of maximum shoulder strength of 191%. These values are summarized below in Table 1. Several of the procedures informally adopted by the technicians to complete the task were also outside established safety standards, for example, maintaining insecure footing, stepping onto adjacent frame members from an elevated work platform, lifting above the shoulder, etc.

Table 1 Lifting equation and shoulder moment estimation tool results

Table 1		
NIOSH revised lifting equation results:		
Value name	Origin	Destination
RWL	10.5 lb.	17.8 lb.
LI	10 times the origin RWL	6 times the destination RWL
University of Utah Simplified Shoulder Moment Estimation tool results:		
Maximum available shoulder moment		743 in-lbf
Worst case moment generated by task (calculated actual)		1420 in-lbf
% of maximum available shoulder moment		191%

DESIGN CONSTRAINTS

Based upon a variety of factors (e.g., furnace design, furnace layout in the fabrication facility, cleanliness issues, etc.), a number of design constraints were determined. The cores rest on a cradle composed of two small wedge-shaped steel tubes that run parallel to the core axis. The cradles contact either side of the core bottom, and run to within 5 inches of either end (see figure 1).

These 5-inch long unsupported ends of the cores are the only places to access the core with an end-effector. In this area, there is too little clearance (the cores are separated vertically by just over 1 inch) to access the core easily and safely with a conventional fork-type lifting device. Therefore, extreme low profile of the attachment component that was to fit between the cores was a requirement. An additional constraint was that the core be solidly attached to the lifting mechanism during transport and placement. This requirement stemmed primarily from the difficulty in manually guiding the core at the height of the two upper core bays and the potentially serious consequences of dropping a core (for example, off the end of a "fork lift" type device).

When placed, the core must be rotationally oriented about its longitudinal axis in the core bay to align the electrical shielding on the side of the core such that it will not interfere with the placement of the furnace framework upon reassembly. The horizontal opening of the enclosure into the core bay, in some cases, is narrower than the length of the core, requiring that the core be placed in the bay one end at a time. Also, a protruding enclosure at the floor level houses temperature control electronics and restricts usable approach space in the maintenance access area.

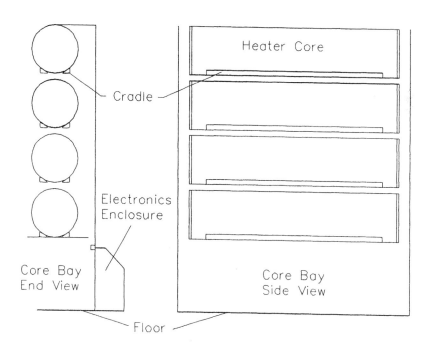

Figure 1
End and Side view schematic of the heater core bay

Additional constraints result from the particular construction and environmental requirements of the cleanroom. An overhead HEPA-filtration system that completely covers the ceiling eliminated the possibility of attaching an upper jib crane pivot or trolley. A suspended metal floor positioned about a foot over a perforated structural concrete ("waffle") floor makes cantilever attachment of a crane to the floor very difficult. Two of the furnace maintenance bays (where the cores reside) face each other and are located on a dedicated isle used to move equipment into and out of the fab, making placement of a permanent jib impossible. These constraints eliminated all but a portable system.

The combined impact of these constraints dictated that the intervention method allow the core to be completely supported by non-manual means throughout its travel from the loading dock and into the fab. In addition, it must be mobile in five degrees of freedom (three translational, two rotational), portable, and keep all forces applied by the technicians within appropriate values. With

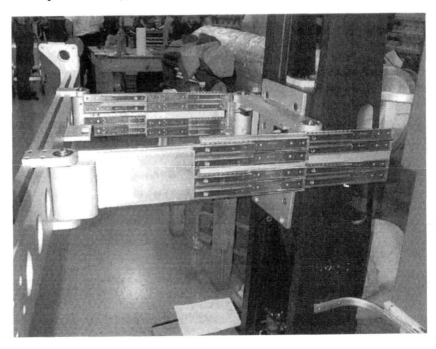

Photo 2
First arm extension concept

these formidable design requirements, there was no off-the-shelf equipment available which could accomplish them.

DESIGN PROCESS

When an extensive search for off-the-shelf portable hoists failed, a search began for companies that construct custom hoists. Alum-A-Lift of Winston, Georgia was selected. Design parameters and constraints were sent to the vendor and their engineer visited the Corvallis HP site for measuring and documenting the problem task and constraints. The cooperative design effort required several iterations and the prototype required one major redesign/rebuild. The prototype was initially constructed using a concept that employed several stacked industrial drawer slides (see Photo 2). This design failed tests at the vendor's facility when HP's inspectors could not move the core through the required range of motion within the specified user input force requirements. Specifically, the drawer slides were unable to withstand the bending loads placed on the arms without severe binding. HP inspectors estimated a required two hundred pounds of user input force to extend the arms the final six inches of travel.

HP's project engineer proposed the original concept of dual articulated arms and the second design iteration returned to this design. However, maintenance aisle access constraints limited the width of the hoist to 36 inches. This resulted in a relatively narrow hoist base. The design required some means to address the problem of excessive side swing of the manipulator. With a heater core in place and the small footprint, there was a risk of tipping the hoist over to the side. Vendor engineers were able to install stops that effectively limited the arm travel while still allowing full extension of the manipulator. In this configuration, it is possible to move the core to the limits of motion in all required degrees of freedom with less than 5 pounds of user input force while maintaining suitable stability of the hoist (see Photo 3).

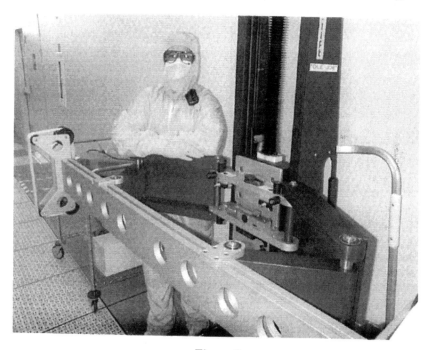

Figure 3
Final configuration of the hoist manipulator

To attach the core to the manipulator, an attachment system was designed that uses two rollers and two pulleys on each end of a 5 ½ foot long end effector with a webbing strap which girds the heater core (see Photo 4).

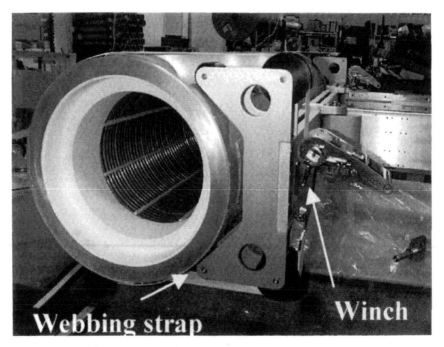

Photo 4
End effector/core attachment detail

ACCESSORIES

To enhance the stability of the hoist when in position to transfer a core to or from the core bay, brakes were requested from the vendor. After several design iterations of the brake, an outrigger mounted, foot actuated and released brake was employed, which works very well.

Two inches of clearance below the furnace and electronics enclosure on all of the furnace stacks allows the use of outriggers that extend under the enclosure. This design prevents the possibility of the hoist tipping forward in the event that an operator inadvertently attempts to lift the core into the framework above while still in its bay. The extension of the outriggers is electrically interlocked with the end effector. The end effector is equipped with a mechanically latched and electrically unlatched catch that holds it in the fully retracted position when stowed. The unlatch function is disabled if the outriggers are not locked in their fully extended positions. If the outriggers are unlocked from their fully extended positions before the end effector is latched in its fully retracted position, a loud alarm horn sounds. This system is very simple but prevents the technician from unknowingly getting into a dangerous tipping situation (see Photo 5).

With the hoist in place for extraction of a heater core, there was no space for a conventional scaffolding or elevated work platform to fit into the bay. Thus, a dedicated platform that could reach over the electronics enclosure was needed. The platform had to be easily moved and set up, easy to use without any danger of tipping, require very little floor space, and provide a cleanroom compatible surface with secure footing. Two platforms are required, one on each side of the hoist. In the final design, handles on each side of the platforms permitted easily pulling them on their wheels. A ½ inch stainless steel plate at the base provides enough counter-mass to keep the platform from tipping during transport and when first stepped onto (see Photo 6).

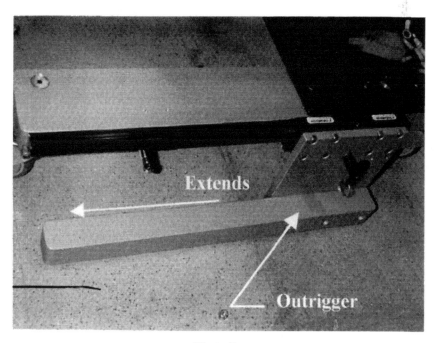

Photo 5
Hoist outrigger detail

POST-PROJECT EVALUATION

The project engineer worked closely with technicians who use the hoist to evaluate the hoist system when first used to replace heater cores. Although there were several minor issues that resulted in changes, such as reorienting handles used to move the end effector, the hoist worked well from the first use. Technicians like the ease with which the cores can be moved and the control that can be exercised when placing the cores. Although the recommended procedure is to have two technicians replace a core, a single technician of any stature can replace the core faster than a crew of four before the hoist was in place, and with much greater control and less jostling of the other process tubes in the stack. This results in less downtime and increased production. Although the benefits to production are clear, the major advantage of the project is the reduced risk to workers. The hoist met or exceeded every design goal of the project with no disadvantages. A survey of the technician's group after the hoist had been in place for several months indicated a high level of satisfaction with the final project resolution.

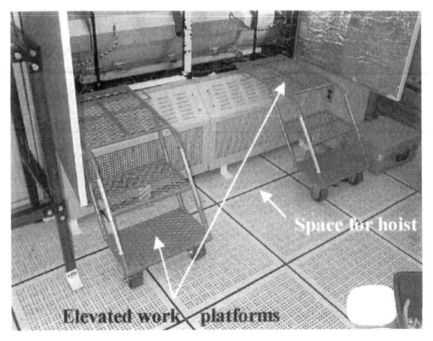

Photo 6
Elevated work platforms in place over electronics housings

ROLE OF THE WORKSITE REDESIGN PROGRAM (WRP)

Near the inception of this project, the project engineer learned of a progressive program through the Oregon division of the Occupational Safety and Health Administration. This program provides funds for the development of engineering resolutions to workplace problems of an ergonomic nature for which no off-the-shelf solution exists. A provision of the program is that the results of the solution be disseminated to industry, especially to industry within the State of Oregon so that other workers may benefit from the project. Hewlett-Packard's policy on the release of information concerning safety and ergonomic improvements is consistent with this provision.

The application process is detailed, but straightforward and the grant requires periodic reports. Up to 90% of the cost of the design/prototyping portion of the project is paid by the WRP to a maximum of $150,000. The cost of this project was about half that amount. Additional grant money is available to implement the solution if more units (hoists) are needed at HP's facilities. This grant is also available to other companies in the industry (must also be within Oregon) that could use this type of hoist. A number of avenues were pursued at the conclusion of the project to make other companies aware of the improvement, but the most productive thus far has been to contact the sales force of the heater core supplier. The primary benefit from the project engineer's perspective is that the project was much less at the mercy of budgetary constraints because of the grant monies and could, therefore, be conducted with greater flexibility to thoroughly address all aspects of the issue. The end result was a completely successful project with no unresolved problems or gaps in the solution.

SUMMARY

Technicians recognized the task of removing and replacing heater cores in the horizontal furnaces to be an undesirable lifting situation long before this project began. Several attempts had been made to find a commercially available solution but were unsuccessful. When a possible connection between this task and an injury was made, the increased urgency drove the project to a rapid resolution. The division Ergonomics Design Engineer at HP recognized the need for a custom solution and Alum-A-Lift was selected as the most viable vendor to provide the hoist. The OR-OSHA Worksite Redesign Program was utilized to provide development funding for the project that began in the spring of 1998. Several design iterations later, the project was concluded in the spring of 1999. The hoist system eliminates the risks of the previous task and substantially decreases the time required to replace a heater core. Maintenance technicians were heavily involved in the design as it progressed, and (perhaps as a result) have expressed great satisfaction with the results.

CHAPTER THIRTY FOUR

Powered Circular Knife to Reduce Worker Fatigue

John E. Johnson, Ph.D., CPE and
Donald E. Wasserman, MSEE, MBA

INTRODUCTION

There are two basic principles that tool designers must consider, tool functionality and operator interface. Equipment is designed to perform certain functions that increase the systems' efficiency. If the equipment does not perform, then it provides little use. Operator interface is a path that has not been as well traveled as the functionality path. In this case study, we will look at the manner in which a powered hand tool was designed to reduce worker fatigue. In order to minimize some risk factors associated with a manual straight knife, a powered rotary knife was designed. By powering the blade of the knife the force required to cut product was significantly reduced. The handle was designed with a lobe shape to significantly minimize forces on the 4th and 5th digits along with utilizing a rubber material, Santoprene®, to reduce soft-tissue compression and damp vibration.

Figure 1: Powered Circular Knife (Whizard®)

BACKGROUND

Power tools should be considered when physical stress is increased with manual hand tool use. Power hand tool selection should consider weight and load distribution, triggers, feed accessories, vibration, and acoustic noise (Radwin, Haney, 1997). By designing a powered tool that replaces manual knives, workers can be healthier and more efficient in their work environment. Many processes that require straight knife cutting introduce several risk factors such as grip force, compression of soft tissues of the hand and muscle fatigue. Two significant problems with using a straight knife to cut material such as foam or meat is that high forces are needed, and handles are usually only available in one size and are not designed for long periods of use (>8 hours).

HANDLE CONTOUR/MATERIAL

There is a natural tendency for operators to grasp onto a tool as tightly as possible when first gripping, then loosen the grip until the tool can be controlled. Through field research in mostly meat plants, it was discovered that the grasping force of tools was directly related to the size of the handle and number of gloves worn. This variability (between and within subject) affects the ability of the designer to adequately accommodate a population in a design (Marras, 1994). Theoretically, equipment should be designed to accommodate the smallest and largest people in the population; however, this is not always feasible. A more common approach is to design for a first, fifth, ninetieth or ninety-fifty percentile (Armstrong, 1995). Since one grip does not fit all workers, a study was conducted to measure the hands of workers. Analysis of the data reflected the proportionality of the digits in relationship to each other across gender and ethnic groups (Johnson, Rapp, 1997). The wide ranges of sizes necessitated the design of four handle sizes.

The contour of the handle was a key area thoroughly researched. From the original design of a conical shaped hand piece, the new design needed to maximize hand surface contact, avoid sharp protrusions and provide an elliptical shape that would aid in control of the tool (resistance to torque) while transferring the work of the hand to the thumb, index and middle fingers. The thumb, index and middle fingers are the strongest fingers, and they should be utilized for producing the most grip force (Radwin, Haney, 1995). Additionally, it was concluded that an optimal handle for maximum strength has a cylindrical or elliptical cross section with an average diameter of 3.75 cm (Chaffin, et al., 1978).

VIBRATION

One law of nature that must be accepted is that all power driven tools emit some level of vibration. The handles or tools cannot be separated or isolated from their power sources. As with any power tool, vibration needs to be controlled and minimized. The basic premise of this circular cutting knife is to change the rotating power to a blade through a unique gear mesh.

In this tool's design, the meshing of gear teeth is a uniquely engineered feature that most would not consider important, but reduction of vibration is important. The pitch and number of teeth of both the drive pinion and the blade evolved through much iteration to provide the minimum reduction in friction and vibration. From transferring one rotary motion into another direction generates some level of vibration acceleration. This area needed to be carefully watched when designing the different sizes of tools.

The table below demonstrates the operating ranges of the tools that have been tested in actual field use. Higher levels were recorded for some tools when observing the population of 685 individual tools. But, as the two columns to the right reflect, once extremely high tools were removed from the population, acceleration levels significantly dropped. What was the contribution to these higher levels in the population?

In this field study, those tools that had high acceleration levels were examined and found to have worn parts or had been poorly maintained. Once the worn parts had been replaced, the tools were tested again and the levels dropped.

Table 1: Tool Diameter and Acceleration Levels (Population = 685)

Tool Diameter (inches)	Standard Deviation (m/s^2)		
	Population	<10m/s^2	<5m/s^2
1.125	0.426	0.426	0.426
2.0	3.895	1.370	0.657
2.5	1.396	1.396	1.042
3.0	1.279	1.279	0.900
5.0	1.004	0.070	0.703

CONCLUSION

Many different areas need to be considered when designing hand tools, whether manual or powered. Of course, powering a tool is intended to reduce the forces required by operators and thusly reducing fatigue and risk factors, but other risks can be introduced at that time. Obviously, vibration is a natural element of powered hand tools. Therefore, it is the manufacturers' moral and ethical responsibility to design tools that have minimum vibration levels.

Simply designing a tool or piece of equipment shouldn't stop at the manufacturer's level. The tool should be introduced to the operator and clearly shown its proper use and maintenance. This would be as if a "perfect" hammer were designed and then tried to be used to turn screws. Again, the operator needs to be considered part of the whole system. But, clearly one of the most important aspects of minimizing vibration levels in powered hand tools is maintenance. Preventative maintenance and replacement of worn parts can ensure that an operator of hand tools is not exposed to high levels of vibration.

REFERENCES

Radwin, R.G., Haney, J.T. 1995. *An Ergonomics Guide to Hand Tools.*

Marris, W.S. 1994. *The Ohio State University Short Course*

Armstrong, T.J. 1995. *Analysis and Design of Jobs for Control of Upper Limb Musculoskeletal Disorders. Hand Tools. Ergonomics Issues in Evaluation and Selection.*

Johnson, J.E., Rapp, G. 1997. Hand Anthropometry in Relation to Hand Tolls and Personal Protective Equipment. In *Proceedings of the 14th Triennial Congress of the IEA.*

Chaffin, D.B., Miodonski, R., Stobbe, T., Boydstun, L., Armstrong, T.J. 1978. An Ergonomic Basis for Recommendations Pertaining to Specific Sections of OSHA Standard 26 CFR Part 1910, Subpart D. *Walking and Working Surfaces*, NTIS Report Number PB 284370, National Technical Information Service. Springfield, VA.

CHAPTER THIRTY FIVE

Use of Three-Dimensional Fabric Yields Ergonomic Benefits in Upholstery Work

Mark E. Benden, MS, CPE

ABSTRACT

Historically, the highest risk of upper extremity muscular skeletal disorder, associated with furniture assembly, has been associated with the upholstery process. In an effort to reduce the risk associated with this group of operators, Neutral Posture Ergonomics, Inc. has developed a new process and product that reduces risks, increases efficiencies and eliminates waste. In addition to the ergonomic benefits found in the actual upholstery work, the addition of three-dimensional (3D) knitted fabric into our product line also offers our customer ergonomic benefits that far exceed the benefits associated with typical materials. This type of process is a natural "win-win" for both manufacturing and marketing.

The study demonstrates the results of both the production floor studies and the end-user studies. The production floor studies will focus on cycle time reduction, force reductions, postural improvements, general cost savings and waste elimination. The end-user study was conducted by the National Science Foundation Industry/University Cooperative Research Center in Ergonomics at Texas A&M University. This study measured the difference (in terms of peak and average pressure) between this new textile and a more conventional fabric that we offer. The equipment used was the X-sensor Pressure Measuring Pad. The result was as follows: there is a 17% decrease in the amount of peak pressure on a person's buttocks and thighs when they sit in the same model chair that is covered in the new 3D knitted fabric. What this means to the end user of a chair is they feel the full effect of our contoured foam and seat-pan and the chair sits more comfortably.

As with any industrial ergonomic improvement, it is beneficial to the mission of ergonomics when both the employee producing the product and the end-consumer using the product can benefit from the enhancements. This study demonstrates the need for ergonomists to consider both aspects of their industrial improvement efforts.

Introduction

Upholstery work for office chairs typically involves cutting of material from bulk rolls, application of the material over foam or padding and then attachment to the wood or plastic understructure of the chair seat and back. In some cases, glue is used to help adhere the material to the foam and reduce wrinkling. For difficult materials (such as leather) or for complex curves and corners, a sewn pattern of the seat is often made prior to assembly to the foam and understructure. While these patterns ease attachment, they add significant labor to the process and may create pressure points under the buttock and thighs when used in the seat pan.

Neutral Posture Ergonomics, Inc. ("NPE") has chosen to forgo the use of seams in favor of a more comfortable and aesthetic final upholstery process known as "pulling". Pulling benefits the user by presenting a seamless, wrinkle-free fabric surface. The puller, on the other hand, is presented with a difficult manual task that involves very forceful use of pinch grips with both hands, large repetitions of this maneuver during the course of their shift and awkward working postures due to various parts requiring a variety of working positions for the actual pulling and stapling of the material.

Common injuries for pullers included carpal tunnel syndrome, tendonitis, trigger finger and low back injuries. For years, NPE has attempted to alleviate the symptoms by building in work breaks, scheduling job rotation, providing proper working height tables and ergonomic staple guns. Despite the minor gains from these improvements, NPE was convinced that outputs from each individual should be limited to 25 chairs per shift to further reduce the risk of injury or illness.

This mandatory injunction against increased productivity, worker pain and suffering and a strong desire to find new ways to delight our end-users was the impetus for our pulling department project. We began with a search of methods to further reduce pulling stress. Several methods including foam compression and sewn seams were rejected due to user complaints or poor finish quality. A semi-automated process was considered as well. This process applies fabric to foam when the foam is molded. High volumes of same color fabrics are required to make this cost effective. It also creates a non-breathable under-barrier that would lead to large amounts of heat buildup during use. The final option we considered was using a material that had been pre-knitted into seat- and back-like shapes that fit our products exactly.

For many reasons, we chose to pursue testing and development of the 3d-knitting option. Just as with a sock or stocking cap, knitted materials have the ability to begin with an odd shape and then conform to another shape by stretching in both the x axis and y axis. Normal woven material, such as that used by most manufacturers is not able to stretch in either direction. Other benefits we discovered included improved heat dissipation (breathable) and exceptional wear resistance when combined with modern yarns that are solution dyed rather than surface dyed.

Illustration 1 - Material Comparison

WOVEN FABRIC   KNITTED FABRIC

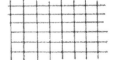   

Each knitted yarn is comprised of a series of double loops that intertwine. The intertwining allows the knitted construction to flex, not the fibers. In the NPE process, the loops on one side are glued but the loops on the top layer are free to move. Three-dimensional knitting is a revolutionary upholstery process. It is revolutionary in that it enables production of a fully shaped chair cover incorporating predetermined patterns in a single stage yarn-to-product process. Using sophisticated computer-aided techniques, 3D-knitted fabrics are designed to follow the curves and contours of the chair cushions, ensuring a smooth fit. Woven thread has a non-flexing basket-weave pattern. When a person sits on woven upholstered cover seating, the fibers stretch which leads to fiber breakdown over time.

The NPE pulling project team was made up of operators (including pullers), industrial engineers, ergonomists, marketing representatives and manufacturing leadership. This group decided to pursue three different tests to determine whether the new material would become our standard. The first test was to conduct a subjective survey of customers and employees to determine whether perceived texture, heat dissipation, cushion and fabric appearance were affected with respect to our standard woven fabrics. The second test was an objective review of the materials performance by the NSF University Cooperative Research Center for Ergonomics at Texas A&M. The third test was a full ergonomic review of the manufacturing process of old fabric vs. new.

A vendor for 3d-knitted materials was selected and custom orders placed for pattern making. Once properly sized patterns were produced, the majority of the tests could begin. Marketing tests related to pattern and color combinations would be performed later if the other tests had favorable results.

Test 1 – Subjective Field Study

Methods

Twenty chairs were pulled, assembled and inspected. Thirty people were asked to review the chairs by using them over an eight-hour shift and provide answers to questions regarding fabric texture, heat dissipation, cushioning support and appearance. Comment sections were provided to allow subjects to elaborate on their answers if they desired. The choices were as follows:

Satisfactory (if satisfactory, are the results better or worse than the standard
material)
Unsatisfactory (would not meet minimum requirements for end user)
No Opinion

Results

Table 1: Subjective Survey Results

	Texture	Heat	Cushion	Appearance
Satisfactory	28	22	21	25
Unsatisfactory	2	3	6	3
No Opinion	0	5	3	2

Table 2: Subjective Results (Better than old/Worse than old)

	Texture	Heat	Cushion	Appearance
Better	28	26	27	26
Worse	2	4	3	4

Discussion

In most cases where we analyzed negative subjective results, we found poor understanding of the test guidelines for subjects. Therefore, with generally good results and a strong positive opinion by the core project team, we proceeded with the manufacturing and objective analysis.

Other results from this test included use of the manufacturing effort as a pilot test for the actual production comparison. This was done to reduce the learning curve necessary to pull this new material and to offer production and quality assurance a chance to develop new methods, if needed, for the manufacturing comparison. The pilot did in fact yield a new method for pulling the fabric. This method was dubbed "placing" to emphasize the difference between the old and new techniques. Additionally, the old process of cutting in one department and delivering to another department for initial gluing to a cushion had to be modified since the fabric department would no longer be needed when using this material. The new material could go straight from inventory to the upholsterer for "placing."

Test 2 – Manufacturing Study

Methods

Two different pullers were timed for start to finish times on pulled 3d-knitted fabric on 10 each 8 backs and 10 each 5 seats with molded foam and one with hand-trimmed memory foam. Two different pullers were timed for start to finish times on pulled Tuff Cloth and Regency on 10 each 8 backs and 10 each 5 seats with molded foam.

Results

Average pull times for the three materials were calculated.

Table 3: Pulling times old vs. new

	Backs	**Seats**	**Total for chair**
3d-knitted	3 min	3.25 min	6.25 min
Tuff Cloth (woven)	3.5 min	4 min	7.5 min
Regency (woven)	4.75 min	5 min	9.75 min

Quality Results:

Quality inspection based on standard finished goods inspection criteria:
- 3D-knitted – 5/40 rejects (rejects were technique related and not attributed to the condition of the raw material)
- Tuff cloth - 0/40 rejects
- Regency - 0/40 rejects

Discussion

3D-knitted material was significantly better in pulling speed for overall chair upholstery time. When combined with the elimination of waste material and the normal time associated with cutting pieces off of rolls for upholstery, the time savings are dramatic.

Subjective feedback from pullers regarding body part discomfort was consistent for the level of physical stress. 3D-knitted material was found to be the easiest with Tuff Cloth a close second and Regency a definite third. This correlates with the results from the time study.

Test 3 – Objective Lab Study

The NSF Industry/University Cooperative Research Center for Ergonomics at TAMU conducted a study entitled "Buttock-Thigh Pressure Measurement Results of NPE Chairs and Competitor Chairs". A complete presentation of the study can be found at the center's web site (http://ergo-center.tamu.edu/npe/buttockthigh.htm) or through our home page at www.iGoErgo.com. The study used capacitive pressure mapping equipment to compare our potentially new 3D-knitted material against our chairs and others in standard fabrics. The analysis included statistical calculations based on mean peak pressure, mean average pressure and mean active cells per pad. Normality, ANOVA and Tukey's tests were used to analyze the data. In summary, the chairs using 3D-knitted fabric were found to be different from the other chairs using standard materials even when standard materials were used with equivalent seat pans and foam.

Table 4: Pressure Mapping Results

Female: Subject 1 Peak Pressure = 205 mm Hg Competitor standard woven material	Female: Subject 1 Peak Pressure = 118 mm Hg NPE 8700 3d-knitted material

Pressure maps such as those shown here are used to determine the amount of pressure felt by the buttock-thigh area while seated in various seats. The measurement device uses 1300 capacitive sensors in a flexible grid placed between the user and the chair. This grid feeds backpressure measurements in millimeters of mercury, which are then translated into a color-coded graph, with blue indicating the lowest and red indicating the highest.

Conclusion

Combining the positive results of the three tests, it is easy to see why NPE approved 3D-knitted fabric as the standard for its top grade of chairs. For the core team, the project represented a rewarding achievement for all members by being able to present a new product that would enhance consumer comfort and demand while improving plant efficiency and reducing the ergonomic risk associated with upholstery work. To bring closure to the process, the final act of the project team was to name the material. "Cloud 9" seemed to fit perfectly since the material left one with the feeling that you were "floating on a cloud".

During the six months since its initial introduction, the material has been a complete success. Manufacturing improvements held true while recordable injuries and illnesses have been eliminated. This certainly dispels wide held beliefs that ergonomic improvements can be detrimental to production outputs or that they will be unacceptably expensive. In fact, the economics of the material continue to improve as techniques mature and purchasing demand creates price breaks. As a tribute to the success of the material, most dealers and end-users that have sampled the product refuse to use or sell anything else on our chairs! This project is one more testament of how marketing, manufacturing and product development can utilize ergonomic initiatives to fuel company growth.

Author Index

Keyword Index

Printed and bound by CPI Group (UK) Ltd, Croydon, CR0 4YY

23/10/2024

01778238-0010